BACKYARD PIG FARMING

backyard pig farming

by

ann williams

Published 1978 by

PRISM PRESS
Stable Court,
Chalmington,
Dorchester,
Dorset DT2 0HB

ISBN 0 904727 69 6 Hardback
ISBN 0 904727 70 X Paperback

Reprinted by the Bowering Press, London

contents

acknowledgements

This book owes much to Colin Spooner of Prism Press who suggested it and helped it along. It also owes a lot to the pigs our family has kept, of all shapes, colours and sizes – bless them all. My parents and brothers have made useful suggestions and helped me to find some of the illustrations and the staff at Ripon Library have been ready with books and references as always.

Ann Williams

MORNING

1 why keep pigs?

Anyone with an urge to produce their own food should be happy keeping pigs. Here are some of the advantages:

- You will be producing your own pork and bacon to suit your family's needs.
- Independence from the commercial meat industry with fewer butchers' bills.
- Your waste food is used up and converted into meat.
- The pigs will produce good manure for the garden and can be used to cultivate it as well.
- The interest and involvement with a fascinating and friendly animal.

There are some drawbacks which should be considered:

- There is a financial risk, as with any livestock enterprise. Stock can die very easily.
- The variable nature of the profits in the pig trade ('muck or money') makes it hard to budget accurately.
- The responsibility of keeping an animal which will need daily attention.
- The possibility of the smell upsetting your family and neighbours.

Let's look at these points in rather more detail. Firstly, the end product. The meat you will produce will be fresh and wholesome and you may find that it tastes remarkably different from the chainstore variety. You will probably notice a difference in colour; the meat will be darker, less pink than shop meat. This is because a lot of meat is treated with chemicals to make it look more attractive. Then there are tenderisers and the all-pervading preservatives which are added to cooked meat. Large amounts of pigmeat are sold in cooked form.

Your home fed pork will be free from these artificial substances and it will also be natural in another way; it will probably taste better because of your methods of pig keeping, in the same way that backyard hens taste better than broilers. Many pigs produced for slaughter are reared in darkened buildings and on a monotonous diet. Some of them are fed large amounts of dairy products such as whey, which tends to produce flabby and tasteless pork.

Our family noticed this difference a few years ago when we sent off two pigs to be killed for the freezer. There was a mix-up and we got back one of our own and one from somewhere else. We cut them up for freezing without realising at first what had happened, but the difference was soon obvious. The pig which had been reared elsewhere was very pale and the meat was flabby, so that you could pick out joints from each pig as they went into the bags and identify them. The teeth of the flabby pig were bad, a thing which never seemed to happen with our pigs. But then ours were living on a varied diet of swill with a little meal; they got fish and fruit and stale buns. The firm red meat thus produced contrasted strongly with the pale stuff, which, at a guess, was probably 'dairy-fed' pork, fattened in the dark. It may also have been younger than our pig, though at the same weight. Old farmers will tell you that a pig tastes better if it is killed with 'a bit of age on it'.

As to time, a medium pork pig which will give you reasonable family joints will take at least five or six months to produce on the backyard system. Bacon pigs are rather bigger and so will take a little longer, but the traditional cottager's pig was a year old when it was killed and one or two people still keep their pigs a full year. The pigs thus produced get rather too fat by modern standards and the joints are very big.

If your pigs are allowed to roam about outside, they will pick up some of their own food; green vegetation, fruit and anything left lying about is grist to their mill. It may take

longer to produce the meat because this natural diet will not be so efficient as the specialist pig foods; also, your pig will use up energy snuffling about and keeping warm outside instead of sleeping in the dark. But with this system there will be less work for you and if things are managed properly, a much better life for the pigs.

By the time that it is ready to go the porker will weigh about 110 to 149 lb live weight. When killed and on the hook, with the intestines removed, the weight (then called deadweight) will be 80-109 lb. Most of this dead weight will end up in your freezer, but the head and the trotters are included and they are fairly heavy. I will give you some recipes for these later on. There is quite a lot of meat on the head which can be made into brawn if you are willing to persevere.

Then there is bacon, probably a greater money saver than pork at present prices, (remember when calculating costs and benefits that home-produced food means that you need to earn less to survive and therefore you also save tax). Curing your own bacon is a reasonably simple and very interesting process. It may not be too easy to get the finished product exactly to your taste at first, but the principle is simple . It is probably true to say that your taste will change and in time you will come to despise factory-cured bacon, while at the same time the experience will cause your technique to improve and then you can vary the cure to suit your own taste.

The second advantage we listed was independence from the system and all its works. Pig keeping is one of the ways of sidestepping the money treadmill to a certain extent. Eggs and dairy produce may already be supplying much of your family's protein and the variety of meat you get from the pig can complete your diet and really raise your standard of living.

Then the savings; the precise value of the harvest is difficult to predict because prices are always changing, but if you are reasonably efficient there should be a satisfactory balance at the end of the year. Even at times when the commercial pig rearer finds fattening pigs for pork to sell to a butcher to be hardly worthwhile, your backyard pig will be economically justified, because you will be getting not only the farmer's profit, if any, but also the butcher's as well. Both these people make a living from your bought meat before it reaches you, and there will probably be a wholesaler in the middle too. So, in spite of the ups and downs of a notoriously variable trade you are bound to find it worth doing.

It is very satisfying to turn scraps and garden waste into meat. Most households have a certain amount of waste food, potato peelings are usually available and even if you are very economical there will be some scraps left over from family meals. Nearly all this can be made into meat by a pig, and this is the foundation of the cottager's pig tradition.

The cost of feeding the pigs can be cut by mixing meal with these scraps, by feeding garden waste such as weeds and surplus or failed vegetables, or by growing potatoes or other root crops specially for them. Up to half the meal ration can be replaced by potatoes - about 4 lb equals 1 lb barley meal. Potatoes can be fed raw; ideally, you should let the pigs turn over the potato patch and dig up the ones that have been missed after the crop has been harvested. There are always a few!

Cooked potatoes may be more bother as a pig food, but they

are more economical because the pigs get more goodness out of them. We had a steamer once in which we cooked potatoes and the fire underneath it was kept burning by drip-fed sump oil, which we obtained free from garages.

Other people's waste food, or swill, is the real saver. However if you go round collecting it this takes time and also it is out of favour at the moment. At the time of writing there is trouble in the Midlands over some hospitals which have decided to grind up their swill, in the name of hygiene, rather than let it go to the local pig farmers who have all the necessary equipment to boil it and recycle it through pigs. It seems that only in wartime do we get our priorities right!

The reason why I said that swill is out of favour is that it was blamed for the long and expensive outbreak of SVD (Swine Vesicular Disease) from which we are just emerging. There has been a law for a long time stating that swill should be boiled for one hour before being fed to stock, and this is being enforced more strictly than in the past. There are also additional regulations concerning the premises in which swill is handled. However, if you progress to keeping a few sows, it could be worth your while to get a steamer and study the regulations. They are sensible enough, and based on the need to keep raw and cooked swill separate to prevent recontamination. The real danger lies in the bones of the meat content of this leftover food. It is thought that imported meat can bring with it the virus responsible for foot-and-mouth and the newer SVD. Anyone who saw what happened during the foot-and-mouth outbreak of 1967 will be glad to do anything to prevent its happening again. Animals were slaughtered in their thousands and some areas of the country were almost bare of livestock before the outbreak was over.

Fortunately, the country has been free of foot-and-mouth disease since then and we are now virtually free of SVD, so let's hope that this is a problem you won't have to face.

If you think that the value of swill is likely to outweigh the disadvantages, your home area will dictate what you should look for. We have used swill to good effect for our pigs but now we manage without it because of the effort involved. I am sure that the pigs preferred it to the meal they get now, so I try to share out our home-produced waste food to give them a change of diet occasionally. In our case, we were living at the time near a seaside town and so the most plentiful swill consisted of cafe and restaurant leftovers and the bits of fish and potato peelings from chip shops; so this was what we used. If there is a bakery or brewery near your home, or a miller, there may be a different type of waste available. Sometimes a cheese factory may want to dispose of sour milk or bad cheese. Any food-producing premises will have accidents and leftovers and they are usually pleased to dispose of them so long as the stuff is collected regularly. You may be able to feed waste without boiling it if it has never been in contact with meat. Check this with the Local Authority to make sure.

What the pig cannot turn into meat it makes into manure, an asset not to be overlooked. Manures are expensive, gardening is popular and the man with manure is in a good bargaining position! But, look to your own garden first; the straw in the much will add humus to your soil and the improvement will be marked in time.

If your pigs are living outside they will spread their own muck - plus a few parasites, so they will need to be moved on to fresh ground from time to time. The patch of ground they leave behind can be used to grow good crops.

Pigs with rings in their noses will graze quite tidily, but if you want them to break up new ground for you, leave them without rings and they will oblige. The finish will be very uneven, but pigs have even pulled out gorse bushes for us and there is nothing like a sow for pioneer work. However in the wrong place they can wreak havoc and spoil your surroundings, so use the rotavating pig with care.

Perhaps you can count the company as a fringe benefit of keeping pigs. They always have a grunt for you and they remind you quite regularly about feeding times - which can be a drawback in a densely populated area. There is, of course, the occasional mean pig. One sow we had went mean with old age and when she had piglets she would hardly let me inside the pen. There are bursts of bad temper under stress (and do watch out for them!) but in general, pigs are sunny-tempered creatures and relaxing to be with. They can be taken for walks. My grandfather is still remembered as having gone for walks accompanied by a pig and a cat. We have a boar now that I can almost treat like a dog.

But be warned; pigs have acquired a bad name. Mrs. Beeton, that faithful echo of the prejudices of her time, called them 'slovenly and filthy creatures'. Nearer to the mark was the defence of pigs put up by a nineteenth century contributor to Chambers encyclopedia, which pointed out that the prevailing filthiness of pigs was a reflection on their owners. I have found that pigs like to be clean and will keep clean if they can, but some pigs, like some people, are cleverer at it than others.

In fact, you will probably meet some anti-pig prejudice among your neighbours at first. The difficulty could be the smell; later we will look at ways to combat this. A well-timed gift of manure for the garden should do much to help diplomatic relations and it may well remove the complaint by spreading the smell to the complainant! It is surprising how pleased people are to be given muck these days. Once almost replaced by artificial fertilisers, it has come back into fashion again.

What other effort will be involved? There is, of course, the financial risk. Stock costs money; a weaner pig weighing about 55 lb would cost you £17-18 at the time of writing, and they are due to go up again soon. The weaner is the smallest unit you can start with. Then buildings can cost a great deal, or a very little, depending on what you have to start with and how ingenious you are. With the pioneering spirit you can do a great deal with a straw bale hut. The risk in putting money down is really the possibility of losing your stock to some disease which can happen and is very discouraging, to say the least. But pig mortality after the weaning stage is not so high as to make the risk a serious one.

Quite a lot of work can go into a pig enterprise, but once you have everything running smoothly the daily tasks should not be too onerous. After all, looking to the well-being of your animals is enjoyable, not really work. Rather more of a hurdle is the fact that animals like pigs are a tie. They cannot be sent to boarding kennels for the holidays. Will they interfere too much with your way of life? There should perhaps be a family discussion about how the work is to be shared and what is to be done about holidays. One way of solving this problem is to share your pig enterprise with a like-minded family. The cost and the work could be divided, and so could the harvest.

We will now go on to look at pig keeping in more detail; in this book I hope to bring together the best ideas from the traditional ways and from present-day practice. There are bad aspects of both the old and the new ways, for we have no monopoly of cruelty and some of the old traditional ways were barbaric. But, at their best, the time-honoured methods were closer perhaps than we are to Nature and to common sense. We will consider both and hope to learn something from them.

2 pigs past and present

Wild pigs lived in the forests of most of Europe and Asia in prehistoric times, so they must have been used to a wide range of weather conditions in spite of lack of fur to keep them warm. No doubt they had tough skins and a lot of bristles; they were very heavy on the shoulder and had small hams - very unlike the smooth rounded pigs of today.

The females and young ones stayed together for safety and the boars usually wandered off alone. Hunting boar was the

thing to do in Norman times and later, and the game laws protected wild pigs in the tenth and eleventh centuries. Wild pig was reserved for the tables of the rich and one wonders how it must have tasted!

There must have been a great variety of food for these forest pigs; nuts and wild shoots, and insects and small animals that got in the way. Pigs will graze but they are not ruminants, in other words, they do not chew the cud as an aid to digestion. As a matter of fact, they have stomachs like ours, and like us, they prefer variety to a monotonous diet. Wild pigs were energetic creatures, always rooting about. In turning over the soil they would also pick up the minerals they needed to keep themselves healthy.

Modern pigs are both like and unlike the wild pig. They are more likely to rear a large litter than their wild ancestors, having been selected for it by generations of breeding from the best. They are now a different shape entirely and have a much smoother skin. Even so, there is some of the old blood left. Sometimes they revert to the instincts of the wild. With their young, for instance, they can get savagely protective and attack their keepers.

In the wild, the sow made a soft nest of dried grass or bracken in which the young ones were born, which is why a pig can be house trained. It is born knowing not to foul the nest. Little pigs have their eyes open immediately and are running within a minute or two of being born, and at that stage even they will go away into a corner to relieve themselves. One way to tell that a litter is on the way is when a sow starts to make a nest, running about with straw in her mouth. When the nest is arranged to her satisfaction she will lie down in it and then you can expect the piglets to be born.

Pigs must have gone into the deep undergrowth at night when they lived in the forests. They like to burrow deep into something to sleep, which can be expensive if you have to buy them straw. But, given the chance, you can be sure that they will keep the bed very clean, because the instinct to keep the nest clean is connected with the light and pigs go towards the light to dung.

From almost the start of civilisation pigs have been with us, domesticated by our ancestors in patient ways at which we can only guess. Perhaps they threw out scraps to entice the wild pigs nearer to their camps, or perhaps they found a nest and reared the baby pigs, then bred from them. Whichever it was, pigs must have lost most of their fear of mankind a long time ago. However they have never become subservient.

Village life in the Middle Ages would have been much smellier without pigs, strange as this may seem. They were the supreme scavengers. Lean and bristly, with long inquisitive noses, they roamed between the houses eating up all the scraps that people threw out. In the days before dustbins, rubbish was thrown out of the door. Pigs must, in this way, have prevented plagues of flies and perhaps some epidemics. Though unaware of all this, their owners valued the pig for its meat, and a pig was small enough for a poor man to own. Although in the Scottish Highlands it seems that for some obscure racial reason the pig was not popular.

In medieval times when the open field system of farming as a community operated in many villages, the pigs were herded in the woods in summer and allowed to roam over the fields after the harvest had been gathered, to pick up what was

FEEDING PIGS.

left. As the peasant gradually came to have less land owing to the enclosures, he often had to give up his cow. But he kept his pig, in a sty at the bottom of his garden; and so pigs went indoors and got lazier and fatter.

In the eighteenth century, farmers still kept pigs outside in places; but tenancy agreements often stipulated that the pigs had to be rung to prevent rooting up the pastures. This was the age of the Fat Pig and the Enormous Ox, when animals were fattened to stupendous size and then exhibited at shows. Workers in the fields could no doubt eat quantities of fat meat, but even so, those enormous pigs must have caused a great deal of indigestion. By the look of them, they must have suffered from it themselves. One Cheshire pig was recorded in 1774 as weighing 12½ cwt! Waste household scraps fed those pigs, in the days of suet puddings and large families. Plus, of course, garden waste, tail corn and brewing leftovers. Nearly all villagers kept a pig in this long tradition until well within living memory and the idea was revived, along with a great deal of self-sufficiency, during the last war. One of the rules of village life was that they kept a slop pail for the pig, and into this went the washing-up water, along with everything else. They knew all about recycling; and if you can manage without detergents for washing-up perhaps you could follow their example. However, never, never let the pigs have water with detergent in it. Some of the older people can remember these details and they are worth listening to if you can start them off on the subject of pig keeping.

Nobody asked what breed a pig was in the old days. It might be described as a 'Yorkshire sort' in the north. Some of this sort went to America and there our Large White is still called the Yorkshire. Various kinds and colours developed in the different regions and those who are now trying to study the history of the breeds are having a fine time sorting them out.

The nineteenth century was the time of the establishment of breed societies for most of our farm animals. Up to twenty years ago the breeds, having been established, were quite distinct. Then the decided preference for white pigs caused them to push the coloured breeds out, almost to extinction in some cases, until the recent revival of interest in minority breeds.

Even some of the white breeds have been swallowed up by the passion for uniformity which developed with the pig processing factories. There was no room left for the Middle White, a short stumpy pig with an upturned nose, and the Small White

Did they really do this?

seems to have disappeared. The Large White took over, followed by the Landrace after being imported. The Welsh pig seemed to stay in favour by virtue of looking rather like a Landrace. The Welsh pig has been around for a long time and, along with

its cousin the British Lop, is thought to be Celtic and to have been living in the West of Britain ever since the Celts have been there. Some of the white pigs were reported to be the result of a far-off importation of pigs from China and also from Italy - the Neapolitan pig is quoted in some of the old books. These pigs may have softened the angular lines of our native breeds and helped them to become the round pigs we have today.

Coloured breeds included the Large Black, Berkshire, Saddleback, Tamworth (a lovely ginger pig) and Gloucester Old Spot. These breeds, I am glad to say, are making a comeback in the hands of enthusiasts and it should be possible to obtain young pigs of these breeds if you are interested.

The modern trend in commercial pig breeding is towards hybrids, as with poultry. The white breeds have been used as foundation stock because of the continued meat trade resistance to coloured pigs. Nobody liked black bacon rind and there was also the dreaded seedycut, when the black colour was more than skin deep and went into the meat along the belly. So with black sows people tended to use a white boar and thus get the blue and white cross, a hardy fattening pig. But this was not the true hybrid, although it was one really. The modern hybrid is a strain of pig which has been developed in the white breeds for fast fattening and lean meat. They are commercial pigs, sold by large companies. For example, the Camborough is a hybrid which is almost considered a breed, although marketed by one organisation. Hybrids are very efficient at producing meat although they can be rather highly-strung and may be totally unused to going outside.

In fact, specialisation has overtaken pigs as it has other areas of farming, which is a pity. Pigs fit in so well as part of a mixed farm because they can be given mistakes and leftovers which would otherwise go to waste. On a farm there is often unsaleable milk, corn or roots. But pigs are rarely kept as scavengers now, except by backyarders. Economics have forced those who farm for a living to keep more and more animals. The powers that be have encouraged specialisation, just as they have discouraged self-sufficiency and initiative. There is a real rat race in the countryside, the more deadly because it is playing about with the environment. Twenty years ago, families made a living from a small herd of cows, some hens and a few pigs. They cannot do it now. Everybody suffers, even the consumers, but most of all, the animals. They have to be crowded together, kept inside for more of the time and weaned earlier.

Part of the reason for the way things have gone may be the prevailing greed for more money, but some of the blame must fall on the creators of cheap food policies who do not care how that food is produced. Livestock farmers do not, in general, make as much money as people investing the same capital in other businesses. Many farmers are there because they want to live in the country and work with animals. Many would prefer the more natural methods, and keep fewer animals, but they cannot afford to do it. Prices are fixed at the 'factory farming' levels where a lot of meat or milk is produced as cheaply as possible. If farmers choose to reject factory methods they still have to compete with them. But - as a backyarder you can sidestep the whole sorry business; stop subsidising the system and keep your own pigs.

Very few pigs are kept in deep filth any more; modern housing tends towards the clinical. But they are crowded together in dark fattening houses, sometimes in total darkness, where there is nothing to do but sleep and eat. The rations are calculated and weighed to the ounce, and are based on barley meal with secret additives; the formulae are not disclosed but some people suspect tranquillisers. Antibiotics are fed to prevent a flare-up of disease.

One of the more stupid traditions which is still with us is the castration of pigs. This has caused more needless cruelty than necessary. Modern pigs are usually slaughtered before maturity, unless kept for breeding, and therefore there is no point in mutilating them by surgical removal of the testicles. But it is still done almost universally. There was a case for it, I suppose, in the days when a pig took a year or more to fatten. However we could do without it now if it was not for the fact that you cannot sell a bunch of weaner pigs unless the males are castrated (when they become hogs). Nobody wants young boars and butchers will not touch the meat except at rock-bottom prices.

BOAR.

'Boar taint' is the excuse for this prejudice and old boars do smell. I have no doubt that they would be unappetising. The taint is actually caused by a steroid in the carcase fat and it appears when the meat is cooked. They can test for the presence of taint in abbatoirs by applying a hot iron to the carcase. Enough interest in the subject has now been shown for some research to have been carried out on boar meat. The Meat Research Institute did tasting trials and found that most families cannot tell boar meat from hog. In Holland they actually preferred the boar meat.

Since boars have been found to convert their food into meat more efficiently than either hogs or gilts - they show an improvement of nearly 10% - the old argument that they would grow better after castration has been thrown out. They do get fatter as hogs, but this is no longer thought to be desirable.

If you decide not to castrate your pigs, a market will have to be found for your surplus boars. This may be easier in the future because things may change and it could be that castration is eventually made illegal. Meanwhile pigs could be fattened and their meat exchanged for goods and services. This can work quite well; we have a neighbour who borrows our cattle crush and gives us a joint of meat in return.

Having taken a look at the way in which pigs have been kept over the centuries, which system will be best for the backyard pig? We will consider this a little later, but in general, remember that sows are rough and tough, and they fight. Any system has to take account of this unfortunately. Keeping a bunch of sows together in a paddock or yard would be simple if they were amenable. However, the small ones get bullied and the big ones get all the food. And yet they prefer company to being alone. The problem is what to do with the sows between litters. Sow stalls, in which they are confined, are cruel in that the pig is deprived of exercise, but they provide the only system where the sows can lie together with no fighting or bullying. For that reason there is something to be said for them commercially, though not much. They should not be necessary on a backyard scale.

One controversial item which you might find useful is the farrowing crate. In this the sow is confined for a few days after farrowing so as to avoid crushing the piglets before they have learned to get out of her way. The dying wail of a crushed piglet is not pleasant and I have seen these crates save the lives of many baby pigs. The sow is quite happy to be confined for a day or two; she would be lying down most of the time anyway just after giving birth. In a crate, the sow can sit, stand or lie, but she can't turn round. The babies can get to her easily. These things have been denounced as cruel, and if you use one, somebody may tell you it's an 'Iron Maiden', but this is rubbish. Let her out of the crate after three days and she will not suffer at all.

3 before we start

There are one or two problems to face before our first pig is installed, so perhaps we should get them out of the way before we go any further.

The first snag you might meet is smell. Pigs have an intensely piggy smell in the cleanest conditions that you will notice at first, even if you rather like it. Then you will get used to it - but your friends won't. They will know you have been among the pigs unless you take steps to cover your tracks.

This was my biggest worry when we first started keeping pigs. I had to feed them before going to school and I was always afraid that the smell clung to me and was evident in the classroom to the town types. Not only had we piggy smells, but we also fed the pigs on Tottenham Pudding, processed swill produced by the town corporation. It smelt strongly like a cross between silage and Christmas pudding and it stuck to me like glue. It is not so bad when you live and work at home and can make a special effort for visitors, but most of us have to emerge at some time to make a living. So it is best to be prepared.

The best thing to wear is a nylon boiler suit zipped up to the chin, and rubber gloves, when you go among the pigs. A hat may be a good idea as well. These boiler suits are easily washable and they dry quickly, so you can wear a clean one every day if necessary. They do let the smell through to a cert-

ain extent, especially when they are dirty, so it pays to wash them frequently. I find it best to wear old clothes underneath them and to make frequent changes from piggy to non-piggy garments. This is tiresome, but better than smelling of pigs all the time. The rubber gloves are really very useful because it is difficult to get the smell off your skin and they will protect you if a hungry pig takes a bite, as one did to me recently.

There is also the problem of effluent smell sometimes. If the pigs live outside, the muck will disappear unless they are kept too long on one piece of land. When pigs are in sties they need floors with a good slope to the drain. This should be an open channel, taking the liquid outside the building to a proper disposal system; otherwise trouble ensues. Stale pig effluent is disgusting and the neighbours will not put up with it, even if you get used to it. There is also the risk of disease, so it pays to attend to the drains. The solid muck will be taken out of the pen every day with a shovel and the main thing is not to let a large heap accumulate outside the piggery door. Try to barrow it onto the garden every day, where it will be doing some good. Manure left standing in a heap outside loses its value because the plant food gets washed away.

Then there are the buildings to consider. They are a problem in that you need tougher materials than for poultry or even for dairy animals. A sow weighs 300 lbs and she uses her weight to get her own way, which a cow will rarely do. Pigs are also extremely powerful diggers and they will develop a tiny hole in the floor into a major excavation. They will chew woodwork in time, although wooden doors for pens are quite usual. Fortunately, above the height of four feet you can use less durable materials for the walls. Most of the damage that a pig will do is below this height.

The old-fashioned stone or brick pigsty with a yard in front, which often used to be near the back door of the house, handy to receive the slops, was evolved from centuries of study of the habits of the pig and it has never been bettered; so if you have one of these you are well equipped. Pigs like them, provided that there is plenty of straw in the sleeping part, because from the yard the pig can watch things happen and there is nothing a pig likes better, except eating.

I mentioned the straw because there is always the warmth of the building to consider. It is quite easy to spend the winter worrying about whether the pigs are too cold and the summer wondering if they are too hot - in the same building. Insulation must be good, with no condensation. I say this despite reading that the latest type of fattening house aims to keep the pigs in Turkish bath conditions. It helps if the air flow in the building can be adjusted, that is, if doors and windows can be opened and shut according to the wind.

Another problem is the famous pig cycle, the fluctuation in price and therefore profitability. You are bound to be advised not to keep pigs on account of this uncertainty. The trouble seems to be that supply and demand can never stay in equilibrium for long. Perhaps pigs breed too quickly; it takes a short time to increase the number of breeding sows in the country and the government is committed to buying a lot of pigmeat from abroad. So at times there is too much. If the price of pork and bacon goes up, producers are encouraged to keep more pigs and very soon there is over production. The price then goes down and the pig keepers cut down their pigs. This used to happen when the nation's pork was produced by small farmers, and it still seems to happen now that pig production has become agribusiness. It is all further complicated by the variations in the world price of feedingstuffs, which one also suspects can be a political matter rather than a simple one of variable harvests, although these do have an effect.

This is all very depressing for pig keepers, but as a backyarder it should not worry you too much. At whatever point of the cycle you break in, there will be an advantage to be gained. If pigs are down in price it will be cheaper for you to buy your first animals, and who knows, by the time your pork is ready things may have improved.

Probably the worst thing to face about pig keeping is that eventually the pig has to be killed to do you any good. It cannot be milked or sheared. If you really like pigs, the answer is to keep a sow which will be with you for a few years, barring accidents. Her offspring will come and go, but by the time one litter is ready she will have produced some more baby pigs for the children to take an interest in. Once you get used to it, producing your own meat is a satisfying part of self-sufficiency. If you eat meat at all you will have faced the facts of butchering. But in this respect it might be a good idea to send away two pigs together to be killed and ask for one to be sent back - the identity will thus be less certain.

If you have heard about the old-fashioned pig killing, with lots of blood and squeals, this may have put you off the thought of pig keeping. Things have changed since those days, and although no doubt the pig-killing was one of the highlights

KILLING PIGS.

of the country year, we can manage without the drama. The live pig is taken in the trailer to the local abbatoir. The FMC (Fatstock Marketing Corporation) have one near us, and they are willing to accept animals for what they call 'private slaughter'. Since so many farmers have freezers this has become an accepted thing. It is now illegal to kill a pig at home, if you intend to sell any of the meat. A few days later you collect the pig, ready dressed, and if you are willing to pay extra they will usually joint it for you.

In fact, there are not so many problems about keeping pigs if you know what to expect before you start. Once you are used to dealing with animals, the stock sense' you have acquired can be applied to other species. If you have kept poultry or goats, you are half way to being a pig keeper. And if not, the experience with pigs will stand you in good stead for anything else you try.

The Large White

4 choosing your pigs

It could be rather daunting suddenly to find yourself the possessor of a big floppy sow about to have young. There is a certain amount to learn about pigs and it is bound to feel a little strange at first to handle your own pigs after leaning over the wall watching somebody else's. But there are easier ways to start than with a sow; for an introduction you could buy small pigs.

First of all, what breed will be most suitable? It might be a good thing to settle the question before you start looking. On the other hand, a convenient source of pigs near to home might tip the scales in favour of a particular breed. Of course, the 'strain' within a breed is very important and it does pay to get stock which is good of its kind. Scruffy livestock is never a bargain, however cheap it may be.

The various breeds have their good and their bad points. The Large White is an easy pig to sell as there are no colour problems. It is a prolific breed, usually producing a big litter of quick growing piglets. It converts food efficiently; from weaning to 200 lb a Large White will eat less than 5 cwt of meal or its equivalent. Large Whites have been developed through selection so that they will usually grade well (put on flesh in the right places so that they get good marks when judged as a carcase). This can be important if you want to sell pork or bacon pigs, because the price differs according to the

P - C

grade and it can therefore affect the profit.

There are several disadvantages to this breed. They are big pigs and can grow to maturity until they are too large for comfort; a very big boar, for example, is a nuisance because he gets too large for the females to bear his weight. Large White sows can get up to 450 lbs plus, at which size they take up a great deal of room and need a lot of food just to keep them going. There seem to be fewer of these giants about these days so perhaps the tendency with this breed is to smaller pigs.

Even so, a Large White is sometimes a rough pig. It will jump up at the slightest sound and sometimes squash piglets through sheer clumsiness. It will rear up on its hind legs, with the forelegs over the pen wall, to watch what goes on. The Large White takes a great interest in everything. Some people blame the ears! They are 'prick' in contrast to the floppy ears of the more docile breeds - the Large White can, and does, see everything.

The Landrace is another white breed, this time with flop ears. It originated from Sweden; there is also a Danish Landrace but the Danes have refused to export any in case we start to produce Danish bacon. The Landrace is docile, quiet and good with piglets. It is immensely long and lean, suitable for today's taste in lean meat and bacon.

The disadvantage with this breed is a certain weakness in some strains. Some types of Landrace are not very hardy and this could make them unsuitable for backyarding, where you really need a tough, happy, adaptable pig. On the other hand, I saw one recently, grazing outside at about a thousand feet above sea level and looking perfectly happy.

Again, some strains have been found to have weak backs. They are supposed as a breed to possess an extra rib; certainly they are very long, and long people tend to have weak backs. Both these drawbacks are actually the result of 'Landrace fever' which afflicted the people, not the pigs. The first importation from Sweden in 1953 created great excitement.Landrace pigs of any kind made such high prices at sales that they were all kept for breeding, whether suitable or not. There was not sufficient selection and the breed was weakened as a result.

The Welsh was kept pure as a breed until about twenty years ago, when under supervision some Landrace and Large White blood was admitted to pedigree lines to ginger them up. The result is now what is called the Improved Welsh. This is the very old- lop-eared pig of Western Britain; a good mother and a sweet companion, a good little pig which is very easy to handle. Being small, it will need less food for maintenance than a Large White.

Saddlebacks. There used to be a Wessex Saddleback originating in Dorset; it had black back legs.The other one was the Essex Saddleback, very similar but for its white back legs. This latter pig was supposed to be descended from the old English pig , but the two breeds have now merged into one, which is a pity if they have different origins. This is the grazing pig, a really outdoor type, smallish, docile, those flop ears again and a good mother. If you can find them they are good backyard pigs, apart from the problem of colour prejudice. The piglets when bred pure are irresistible, but there may not be as many piglets as say a Large White would produce. But under rugged conditions this may not be a bad thing; it is better to have eight or nine fat little pigs than a large litter of weaklings.

The Saddleback

The Gloucester Old Spot

If you are thinking of keeping pigs outside then the old breeds may be worth a consideration. The Gloucester Old Spot is available again and this used to be the Swill Pig; the description of this breed in a book of mine is headed 'Good Results on Swill Alone'. These again are most attractive pigs and I am glad that they have been saved from extinction. People are beginning to collect test results under commercial conditions for the old breeds, and while I have seen a few random reports which have been quite good, in general I think it is fair to say that one would expect hardiness on the one hand to be offset by smaller litters and perhaps poorer grading.

Then there are the Hybrids to consider. They are sold by very large companies breeding pigs behind locked gates, ostensibly because of disease. So, if you bought a Hybrid, you would probably not be able to choose your pig by the expression on its face; your choice would have to made on paper and according to a mass of figures. You would be practically certain to get a fair deal from these sort of people, but perhaps there would be more human contact and after-sales service from a smaller breeder.

An exception could be the Hybrid pig produced by Mr. Ferguson of New Pitsligo, Aberdeen,who seems to be a real enthusiast. This pig, it is claimed, has been produced 'with wide environmental tolerance' and on their own farm the dry sows go out to grass. I have not dealt with this firm but I have spoken to Mr. Ferguson and he is interested in all aspects of pigs, not just the purely commercial ones.

The advantage of Hybrids in general is a proven ability to fatten fast and economically on conventional pig rations. They have the figures to prove it. We bought a Hybrid boar once and he arrived with a sheaf of papers covered in figures, which could not disguise the fact that he was very ugly indeed. However, it proved to be a lovable ugliness and he did well for several years, but succumbed to old age rather early.

The disadvantages, apart from possible ugliness, of these pigs may be poor nerves and the accompanying probability of a short life and a fast one. I cannot be sure how they would fare under backyard conditions.

There is an organisation which will help you if you have any difficulty in finding the sort of pigs you want. The National Pig Breeders' Association,49 Clarendon Road, Watford, Herts. looks after all the breeds except the Landrace, including the rare breeds. But the pedigree of an animal is not really so important to either its comfort or yours, so do not pay too dearly to be able to trace the ancestry of your animals.

There are a few exotics about in the pig world now, imported during the last few years. If you see something you cannot identify it could be the Pietrain from Belgium or the Duroc from America.

The breed you eventually select depends on you. I should say the Saddleback is the ideal backyard pig, but you may not have a good local supply. Then, with this breed, the colour may be a drawback, but not if you swap meat rather than trying to sell pigs to the butcher. Many people with Saddleback sows overcome this by using a white boar; the resulting piglets are white with grey patches, the 'blue and white' pigs I mentioned before. We had a blue and white cross-bred sow once and she was wonderful. Then we started to sell pigs to a 'weaner group' which had been set up to help small producers with their marketing. The advisor for this group came along and looked down his nose at our spotted sow, but happily, we kept her, and parted from the man.

Supposing the breed and the source of supply have been settled. What sort of pigs will be best to start with? For a safe start I would recommend the way that we began pig keeping. We bought four little pigs from the same litter, a few days

after they had been weaned from the sow. They weighed perhaps 40 lb. If four is too many, buy two, but try not to keep one pig on its own because they really do grow better with a little competition and they appreciate company.

After this you can do one of several things. All the pigs can be reared to pork or bacon weight, disposing of, say, three to the butcher or friends, and retaining one for home use. Or a female can be kept for breeding; at seven or eight months when the others are ready for bacon, she will be big enough to go to the boar, and four months later will produce a litter of her own. This means that you can breed your own pigs within a year of starting.

Fattening bought-in pigs will be more or less profitable according to the state of the pig market, and also, according to how you can cheapen the feed costs. The cost of the weaners comes into it, but never buy scruffy stock because it is cheap. You can reckon a rough £20 cost for each weaner, plus about the same for the cost of food, which gives a total outlay of £40 per pig. For this money about 40lb weight of bacon could be bought at present prices, whereas there will be about 80lb of meat on the pig.This is a very rough guide, but a more detailed one would probably be out of date before this page was printed.

During the time that your first pigs are growing you will be gaining valuable experience. Only by doing it yourself will you find out exactly how much food to give them and also how to compromise between the pigs' needs and your environment. You will find out whether your conditions are suitable for the system of keeping pigs you have chosen. I hope you will find that pig keeping is a satisfying hobby.

By the time that the pigs are ready for pork and bacon, you may feel quite confident about handling them and ready to try breeding your own. There are several advantages in keeping one of the gilts (young female pigs). One is that she will cost you less than a bought-in gilt ready for breeding, which could be up to £80 or more, depending on her pedigree. Also, she will have lived long enough on the premises to have got used to to the bacterial flora there; she will be at home, will know you and you will know her. Make sure that she is a good specimen and has at least twelve teats. If there is a choice, pick the pig which is most friendly, all other things being equal.

There is a disadvantage of breeding from a gilt at the start of your pig-keeping career. Gilts are sometimes nervous and behave badly with the first litter; and it will be your first litter as well! If you are both nervous, things could be tricky.

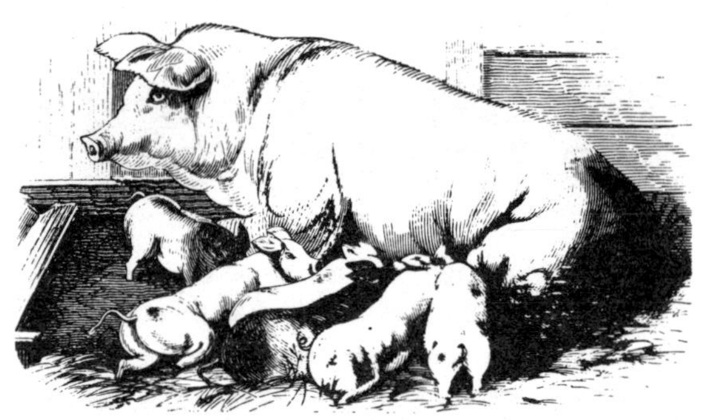

How can this be avoided? I can tell you what to do when a litter is born (see Chapter 7). You could sit by the pig with a lucky charm in one hand and this book in the other. There is always a chance that the pig could have the commonsense to get down to the job and do it properly. Or, you could call in the vet at the first signs of labour, just to be on the safe side. This would add to the cost of course.

A few months ago we had a batch of four gilts farrowed down for the first time. Two behaved perfectly, the other two panicked. They killed the first piglet that arrived and would probably have murdered the lot without interference. This month, three more farrowed and quite naturally I was apprehensive; but they were all perfect little mothers from the start. You just never know what will happen that first time. After this they know what to expect and very rarely treat the piglets badly, though they may get rather annoyed with you.

If you feel at all nervous about farrowing a gilt, sell all the first lot of pigs and look around for a suitable sow, one that knows the ropes but is not too old to rear a large litter.

In my opinion the market is not a good place to buy animals. For one thing, if there is any disease about this is where it spreads. There is also the anonymity which means that you have no contact with the previous owner and little idea of why the animal is being sold, although a very good reason might appear when you get the pig home. This is not to say that all market pigs are bad bargains; but buying any animal in the middle of its breeding life is not so straightforward as buying young animals or fat ones. You must be sure that there is a genuine reason for the sale, and one which will not be a drawback to you. Bad reasons for selling sows would be small litters, bad temper or ill health.

There are, however, sows to be found which are sold for none of the undesirable reasons. A youngish sow might be obtainable from a pig farmer. The average length of life expected of a sow on a modern commercial pig farm is only six litters, which takes three years. So the usual policy is to replace one third of the sows each year by gilts. Perhaps you could step in on this cycle and rescue a sow due for replacement on the grounds of age only. In your environment,without the stresses of living in a large herd, she could go on to have a few more litters. If you do this, ask for an in-pig sow and this will save looking for a boar, for a few months anyway.

To add it up, then. Firstly, the breed does not matter too much for backyard pigs. Neither does it matter whether you start with big ones or little ones. It is important to try to get your stock from a good 'strain' or family within whichever breed you choose. Although long paper pedigrees cost money and they are not always justifiable, good commercial stock will do as well as pedigree animals as a rule. If you can buy from an enthusiast this will be a valuable contact. Such people will be generous with time and advice and enthusiasts do make life more interesting.

The nearer home you can find pigs the better. We once decided to buy two expensive females from a distant part of the country, selecting them on their family records. The pigs were fairly heavily pregnant, and they travelled by rail in a special heated, ventilated livestock compartment. One had an abortion on arrival and the other produced a very small litter.

When buying a sow or a gilt, the main thing is to give her plenty of time. Before she produces young, she must have settled down in her new surroundings and got used to her new food and new owners. Any pigs which you have already, or might have had recently, will have left unfamiliar bugs to which she has no resistance; and here again a little time is worth a lot.

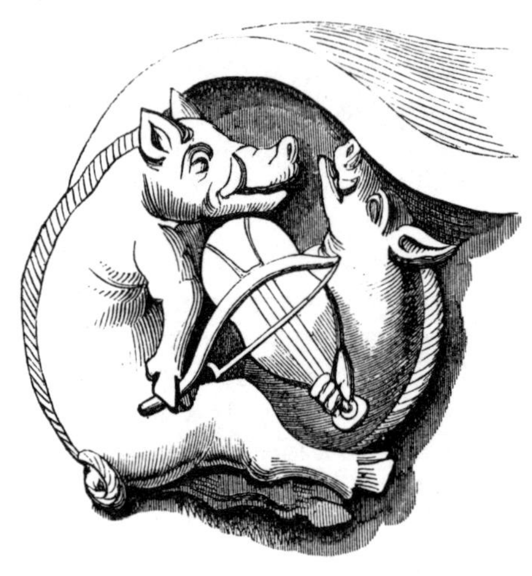

5 housing

Most advice on pig housing starts with a long and expensive list of requirements; but this chapter is not intended to tell you what to do. I thought it would be best to suggest a few alternatives which have worked for various people under different conditions. My main aim is to keep things as simple as possible; we don't want to end up with a small imitation of a factory farm. Fortunes have been, and probably will be, spent on palaces for pigs to live in, but the sophisticated equipment is mostly for the benefit of expensive paid labour. And the whole idea of backyarding is to do the work yourself and to keep agribusiness out.

Firstly then, pigs outside. This is where they would prefer to be and in some ways it is the easiest way to keep pigs. They have a certain amount of choice in what to eat, where to lie and so on, and they are not completely dependent on you for everything if you have given them a good environment to begin with. Nature takes care of such problems as ventilation and mucking out.

I have mentioned before that pigs are rough and tough. They are also marvellous diggers and if you want your land to resemble a tank testing ground, by all means turn them loose. If there is rough scrub land to be reclaimed, pigs will do it better than anything else. However if you only want them to graze and not to move turf, you must ring their noses.

Wet land is obviously unsuitable for pigs, and so is a patch that is too small and gets churned up and foul. Limited access is the answer in either of these cases; keep the pigs inside and let them out for a treat, say at weekends when you can keep an eye on them. Pigs enjoy a gallop so much that it is well worth the trouble of fencing them, to allow them out occasionally.

Types of Housing for Pigs Outside

These are fairly simple. *The corrugated hut* is the ultimate in simplicity; it is sometimes called the Roadnight Shelter, after Richard Roadnight who keeps a great many sows outside successfully. This is simply a curved corrugated roof sitting on the ground, with one end blocked and the other open. Inside is deep straw. Sows farrow in the huts on the Roadnight system; there are three sows to the acre so they have plenty of room and the litters are quite healthy. Litter sizes tend to be smaller as the weaker piglets do not survive, but, in general, they do very well. A simple shelter like this can be moved to another field when the land needs a rest; this is important when you have to move your sow or sows on. If you have just one sow, the best plan is to allow her access to a small paddock and then later resting this by moving her on to fresh ground.

The Straw Bale Hut is a good cheap shelter and it can be burned after being in use for a while. Burning the straw kills off all the bacteria residing in it. Some have a straw roof but the tendency now is to make the hut two bales high and roof it with corrugated iron sheets. The straw walls can be protected from the pigs, who would otherwise demolish them rapidly, by corrugated sheets. These are held in place by stakes driven into the ground, one on each side of the bale with a wire tie between the stakes; or the straw can be protected with sheep netting. Inside the hut there is a sheet of wire mesh under the corrugated iron roof and the space between the wire mesh and the tin is filled with loose straw, for insulation. With deep straw bedding this makes a snug shelter for the sow and her litter. Sometimes several 'dry' sows share a hut between litters,

but if so make sure that they are all friends. One of them might be forced by the others to sleep outside the hut.

The usual dimensions of a straw bale hut are 7 ft by 7 ft 6 in, and one hut takes about 25 bales of straw, possibly less if sheets are used for the roof. The price of straw varies so much that a costing is not feasible; but you don't need to buy straw. Exchange it for manure - many farmers are pleased to have manure these days, or swop it for some other produce. However it is arranged, try to get it straight from the field at harvest time, or alternatively late in the spring. Midway through a hard winter is the worst time to look for straw.

The Wooden Ark. If you want rather more care taken of your pigs outside, you can get or make a wooden ark. There are lots of designs for these. The triangular shape makes it a strong structure and also provides corners where the little pigs can get away from the sow and the danger of being crushed. Some arks are fitted up for farrowing, with creep rails to protect piglets even more, and a nest for them. The doors are baffled to prevent draughts. The wooden floor makes a warm bed and stops draughts. The height at the apex of these buildings is usually 6 ft and the floor area is about 9 ft by 6 ft.

Fold Units. The great advantage of these is that they are moveable huts. The arks already described can be fitted with runners or wheels and can then be dragged by tractor or horse power, the narrow end on through gateways. But the fold unit is even better as a way of controlling the passage of pigs over the field. The unit consists of a wooden hut with a fenced-in run. When the grass in the run has been eaten down, the whole thing is dragged on to fresh ground. If the unit is moved on every three days this will safeguard against a build-up of worms. It is an extra job to do but the system is about the best for the land and the animals. To give you an idea of the amount of land you would need, if each run is 10 ft by 8 ft and you move over a four acre field, you can move ten units every day for a year

and still use fresh ground each time. The pigs are shut into the huts for the operation and they ride along on the wooden floors.

There is nothing nicer than a well-managed outdoor pig set-up. But things can go wrong. Perhaps you have seen a bad one with sows wading about in deep mud with hairy coats and ribs showing, and little pigs with frost bitten tails? Don't keep pigs like this because they will not be happy. They must also have a warm place to sleep with no draughts.

This is not to say that they should be kept completely away from mud; they love the occasional mud bath. We had a place where the sows used to trip up the water trough in hot weather and make a wallow. They came up absolutely plastered, but then they didn't havc to live in wallow conditions, any more than we live in the bath.

The main thing is to know when enough is enough and to bring the pigs inside if things get too bad outside. This can apply equally in summer as well as in winter. Hot weather can be too much for them. All pigs need shade. The white breeds get sunburned and some pigs get sunstroke. It should be possible to arrange some shade for them, but when the flies are biting they would rather be indoors during the heat of the day. In this kind of weather they prefer to go out in the evening when it is cooler.

Fencing may be the costliest item for outdoor pigs and the only way to avoid it would be to put them into fold units. If you have fields with high, solid stone walls and good strong gates this will confine the pigs; anything short of this will need looking at carefully. Electric fencing will keep in pigs pretty well. They need two strands, one at about 15 ins and the other at 2 ft 3 ins. A special fencer battery supplies enough current to give any animal that touches it a shock, though not a dangerous one. The fencing posts need to be insulated of course. The problem that I have found with this form of fencing is that the pigs develop such a respect for the wire, which they probably cannot see, that even when it is removed they will not cross the place where it has been. It becomes a very difficult problem to get them out of the paddock unless there is a proper gate which has not been electrified. This they will happily use.

An electric fence will not keep in piglets; they will run underneath it. This may not matter if the field as a whole is pig-proof, but if you want to keep baby pigs in one place the only thing for it is pig wire, effective but expensive. It is stapled onto wooden fencing posts and provided it is kept taut, is pig proof.

Indoor Pigs.

It may be that you have no room for outside pigs. The concrete raft is a compromise between in and out which could come in useful during a wet winter. If a patch of concrete about 8 ft by 6 ft is laid down, a pig ark or hut can be drawn up to it and the pig will get the benefit of fresh air without being on the paddock. In this way the cheap outdoor housing can still be used when the land is too wet or has become foul. The concrete can be made into a yard with 3 in by 2 in timber frames clad with second-hand corrugated iron sheets.

Cottage Sties. A refinement of this yard idea is to build a cottage sty, a concrete block hut with the yard at the front; the walls of the yard can also be of blocks if you would prefer a permanent job. The huts can be simple, about 6 ft at the front

sloping to 3 ft at the back, with no windows and just a bag over the doorway to cut down draughts. The door should be at one side rather than in the middle. We made one out of blocks manufactured on our own small concrete block machine, which turns out two blocks at a time. Slow work, but it saved money if you didn't charge for labour.

Such an arrangement should do very well for your porkers. It is often assumed that fattening pigs should be inside. They will certainly take longer to fatten if they can gallop over acres of ground, and if they are outside in winter some of their food will be used to keep them warm. All the same, they are healthier with some air and a house with a run is a good compromise. The yard will be easier to clean out than a pen in a building and so your life will be easier as well.

Perhaps you have an old stable or cowshed which would convert into a pig pen. In this case, a farrowing pen would be useful and it would also hold a litter of weaners or fatteners if necessary. Before you start, try to visit a few other people who keep pigs and see how it is done. You can learn from other people's mistakes as well as from their clever ideas.

A few basic points should be borne in mind. As to size, a sow needs about 100 sq ft for herself and her litter; some designs of pen allow more. The floor should be of smooth concrete but not too smooth or it will be slippery when it has muck on it. If you are laying the floor yourself use a wood float for the finish. It will need a good slope to the drain, and if you are making your own drain it should be an open channel that can be brushed clean. If you lay a floor, insulate it. Many things can be used, egg trays or gin bottles or anything which will provide a layer of air between the rubble and the concrete.

If there is already a good solid floor in your proposed pen, there is no need to dig it up in order to insulate it. Make the pig a wooden sleeping platform instead; a couple of old doors will do very well provided that the pig cannot move them. Research has shown that substituting ½ inch of wood for 1 in of concrete raises the temperature by 21.6 deg F for the pig. This is a great difference and well worth the trouble.

Plenty of straw on the floor is well worth the cost, especially for indoor pigs. It keeps them happy in several ways. It cuts down draughts, insulates the floor, gives them something to burrow into and something to do. Deprived of rooting about outside, the indoor pig loves straw. It will also eat a great deal of it and thus get the necessary roughage in the diet.

Ventilation. Try to see that the pigs are not in a draught, especially the little ones. They can stand lower temperatures if the air is still. Scientists say that you should have 5 sq in of outlet ventilation, to take away the stale air, for every 100 lb of pig - that is, every porker. For inlet ventilation they suggest 15 sq in, which sounds a lot. Inlets should be at least 3 ft 6 in above the pigs so that any draught caused goes over the pigs' heads. Control flaps on inlets are very useful, especially on exposed sites. In bad weather they can be shut up.

Insulation. This is a very important factor to consider when pigs have no deep straw. In a building, the roof is the thing to watch; up to 80% of heat losses in buildings can go through the roof. Good insulation keeps the building warm in winter and cool in summer. It also cuts down on dripping condensation.

Bear in mind that if you can keep the heat in, the pigs will provide the warmth themselves. Seven porkers give off as much heat as a single bar electric fire. Don't lose this heat, conserve it. As with any kind of housing, the problem is how to keep the air warm and fresh. How warm should it be? Sows like a temperature of about 60 deg F best; piglets like 80 deg F, but they can be kept happy in the creep with an infra-red lamp. Fatteners are best at 65 to 70 deg F but groups obviously keep warmer than individuals.

It seems to be more important to have no draughts than for pigs to live at high temperatures. They can apparently thrive at anything between 40 and 85 deg F. Scandinavian experiments[suggest that they can do as well at low temperatures as at higher - given plenty of straw.

How do you insulate? There are various levels of sophistication, like there are in houses. We have stretched wire netting under the roof and filled the space up with straw. Materials can be assessed for insulation by their 'k' values. This is a measure of their thermal conductivity. The 'k' value of a building material is the heat in British Thermal Units (Btu), transmitted in one hour through 1 sq ft of material when the difference in temperature between inner and outer sides is 1 degree. The lower the value, then, the better the insulating qualities of the material.

So - if you have no land at all, your pigs can still be happy. Make them a nice light, warm pen and give them plenty of straw. Throw them in a turf every day to play with. Muck the pens out every day if they are small pens or if they contain a lot of pigs.

6 feeding pigs

Of the cost of producing a pig about 75% goes on food, so if economies have to be made this is the first thing to look at. The problem is that cheaper foods tend to be bulky, and pigs cannot deal with a great amount of bulky food, especially if you want them to grow reasonably quickly. A pig will grow three times as fast as a beef animal, so it needs a good ration. Bulky foods are bound to slow down growth. So when finding cheaper foods you have to balance the cost with the speed of production; the pig will not mind taking its time and as a backyarder you do not have to compete with the conveyor-belt production of a large unit, where pigs have to move on at a certain time to make room for the next batch coming on behind.

There must be hundreds of different concentrated pig foods on the market now. They are sold as meal, pellets or cobs. Broadly speaking there are three main classes of food, but feed companies have their own special names for these three categories. First of all there is 'sow meal' which is for breeding stock - it is fed to boars as well. This is usually barley meal with protein added to bring the total protein percentage up to 16% or so. The protein may be fish meal or soya bean meal.

Then there is the fattening ration to be fed from a weight of about 55 lbs, when the pigs come out of the 'weaner' or baby stage. Fattening meal used to be more or less plain barley meal; it used to be cheaper than sow meal. These days, it costs more because today's streamlined pigs are fed on high-powered rations. Adding to the cost are the 'growth promoters' as they are hopefully called, a list of additives now thought to be essential for the well-being of the fattening pig.

Antibiotics. Some of these are used to promote growth and some to prevent disease. Growth ones include virginiamycin and zinc bacitracin. Some of the others have recently been withdrawn as unsafe. Antibiotics to damp down disease are regularly added to the rations in some places, such are the conditions in factories producing pigmeat. But at least these are not added now until they have been definitely prescribed, which is an improvement. They include such things as penecillin and the tetracyclines.

Copper is added to pig fattening foods. It used to be put in at the rate of over 200g per metric tonne but then 'they' decided that this was unsafe and now 175g is the limit. In the EEC, their maximum dose is 125g and we may fall in with this figure soon. But why put it in at all? Copper has been found in pigs' liver and in land treated with pig muck. It may slightly improve the effect of the food on the pig, but it is a risky substance to play around with and can soon reach a toxic level.

Arsenical Compounds are being outlawed now, but when I started to write this book they were still being added to pig rations. I include them to remind you of what you get with your bought meat. They were added to the meal at the rate of 100g per tonne - and residues had been found in carcasses. So the rule then was not to feed these to pigs within ten days of slaughter.

Hormones were not added to pig feed until recently. But now feeders have found it profitable to add to the feed a combination of synthetic androgen and oestrogen. This may, understandably enough, affect reproduction, so they tell you not to feed it to breeding pigs. This again is not supposed to be fed to the pigs just before they are slaughtered. But what about the consumers' reproduction if somebody makes a mistake? The mind boggles.

The way in which these additives work is not fully understood, but this has not diminished the enthusiasm for them. It has been found that they have more effect where the organisation on the farm is wrong, with poor housing, disease and so on. One would think it would be better to put these things right instead of poisoning the pigs- and possibly the people.

The third type of pig food you will find is 'creep feed' for baby pigs, usually in the form of little pellets. There is a very expensive starter creep and sometimes a cheaper one to give them at about one month old. The creep pellets fed to baby pigs can cost up to £250 per tonne. Pigs eat very little at first, but it is a good idea to get them going on solid food as soon as possible, especially with a large litter. With a litter of eight or less, the sow will have enough milk to go round and therefore it will not be so urgent.

I put two churn lids from old milk cans into the creep area when the piglets are four or five days old, one containing

creep feed and the other, water. The water is very important and it is surprising how much baby pigs will drink.

It is easier to buy creep than make it, expensive though it is. Up to five weeks of age each little pig will eat about four lbs of creep, so one bag will do for a litter of twelve. A typical creep ration would include barley, wheat, maize, milk powder and molasses. They love anything sweet and if all else fails, you can start them eating with sugar-coated breakfast cereal. Some litters seem to start earlier than others.

These then are the conventional pig foods, fed in two feeds morning and evening.

Methods of Feeding. I mentioned earlier that hungry pigs are rough and should ideally be fed from outside the pen, so that you need not go among them until their heads are in the trough. I have been knocked over several times by pigs trying to get the feed bucket out of my hand, so if I need to go in with the feed I now throw a handful over the door first to distract their attention. Pigs are also rough with each other at feeding time and thus it is vital that each pig has a place at the trough, unless they are on 'ad lib' feeding, in which case there will be food in a hopper all the time and they can eat whenever they choose. Little pigs that have just been weaned are used to frequent meals and so they should have enough food to allow them to eat frequently, but keep an eye on the amount they eat and see that they do not get too much or they will throw it about and waste it. It is a good plan to let them clear up completely every few days.

Adult pigs can be fed swill and/or meal in several ways. They can be fed dry, and help themselves to water from a separate trough. But this is a wasteful method for outdoor

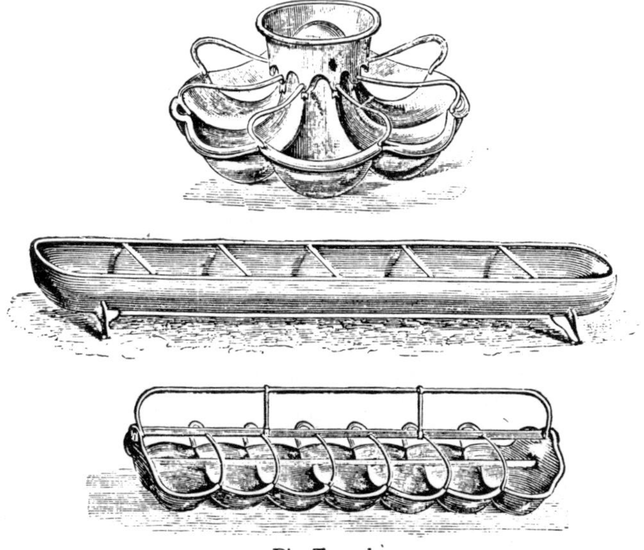

Pig Troughs.

pigs if you use meal- it can be blown away or lost in the grass. The best way to feed dry meal is in a hopper.

Another way is to give them dry meal with water poured on it in the trough- or, as I do it, fill the trough with water after each feed and at mealtimes, measure out the meal in a scoop into the water remaining in the trough. This makes a sloppy mash and most pigs love it.

If you are feeding cooked swill, with or without meal, get it ready in a bucket first. Old books tell you that proper pig food is soaked in a tub for hours before the feed, but this is not done now. Swelling it up with water may have made the pig feel very full, but it wouldn't make the food any more nutritious. Make sure that the swill is really cool before it is fed. No doubt you will have sorted it before cooking to make sure that it contains no foreign objects- we were always made to go through it again when cooked! I once found a sixpence at the bottom of a trough, so the pigs were more efficient sorters than we were.

A simple way to feed pigs outdoors is to throw them a few cubes or cobs. This is meal compressed by the merchant into squares of various sizes; the big cobs may be two or three inches across and are easier to find in the grass. They cost more or less the same as meal and if the pigs are getting grass or vegetables for some of the ration, you could feed cobs once a day.

Floor Feeding. If the pigs are indoors or in yards you could try this method. If the pen is big enough you can scatter the food on part of the floor and the pigs will keep this part very clean. It saves the expense of a trough and ensures a fair share for all because the food is scattered.

Constituents of Food. The food we have been dispensing, of course, consists like our own of the following six compounds: water, protein, carbohydrates, fats, minerals and vitamins. We don't usually know precisely how much of each we need or how much we eat, but try to balance our diet and watch our condition. You can do the same with pigs. Nutrition is a very advanced science and there are (to me) almost incomprehensible books on the subject of feeding pigs, but once swill comes in, science goes out of the window. It is impossible

to do anything but guess at the value of swill and you have to play it by ear. Protein is the most expensive and offal from meat shops, though unattractive, would be the most valuable. Bakery waste will need some protein to balance it. If you are a complete beginner you can always buy ready-made pig meal and feed it according to the instructions for the first month or two, until you get some idea of what a normal pig looks like and eats like. On this basis you can play your own variations later.

Proteins. The protein part of any diet is usually the most expensive and difficult to find. Proteins are composed of several amino-acids, and the quality of the protein depends on how many and which of these are available. Cereal proteins are deficient in the amino-acids lysine and methionine so these are the ones needed in a food to balance a cereal ration whether it is barley meal or stale crusts.

Fish meal has been in the past the most usual protein supplement to cereal diets for pigs because it contains these essential amino-acids and is also a good source of calcium, phosphorus, riboflavin and vitamin B12. Lately fish meal has been replaced by soya bean meal in many feeds because of the high price of fish. There is also the risk of a fishy taint when fish meal, particularly herring meal, is fed in the later stages of fattening. On the other hand, meat and bone meal protein is not so good as fish.

Separated milk is high quality protein when fed in liquid form as it is when home produced, but not so good when it is a powder, dried at high temperatures. It does seem to improve the digestibility of other foods.

Groundnut is a poor source of lysine and methionine and can sometimes contain a poison called aflatoxin. Cotton seed can also be toxic to pigs, as it sometimes contains gossypol.

Protein Supplements. These are sold as meal or pellets for mixing with cereal to make a balanced ration. They consist of a variety of protein and can also include vitamins.

Vitamins. Bought-in pig rations are usually supplemented with extra vitamins A and D. The recommended amounts are 4 million International units of A and 1 million of D per ton. These will only be needed by indoor pigs; pigs on grass can find their own vitamins. If you would like to add these vitamins to a home made ration, the easiest way is to add fresh cod liver oil to the feed. Buy this in small quantities as it does not stay fresh for long.

Riboflavin, one of the B vitamins, is the only other one likely to be short in pig rations. Dried yeast is sometimes used to supply this.

Minerals. Outdoor pigs again will find these in the soil, particularly if they are allowed to dig. For indoor pigs the turf thrown in should put things right except on mineral deficient soils.

Digestion. Pig digestion is similar to human digestion as we said; the intestine is 60ft long in an adult pig.

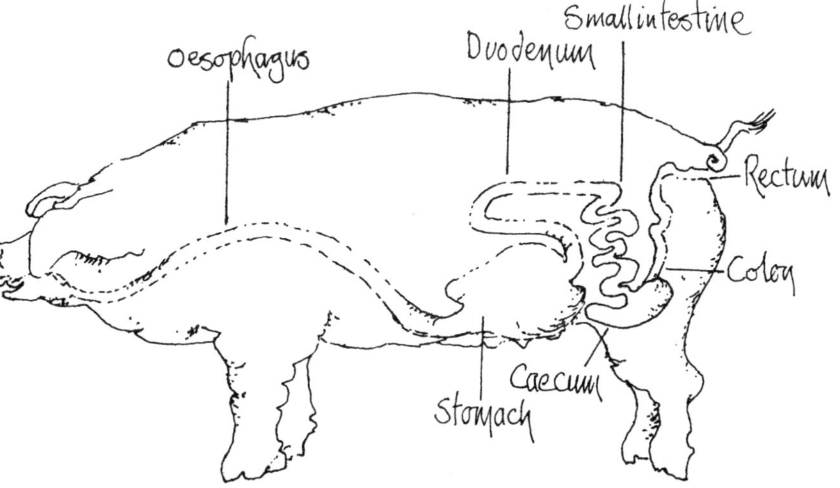

During digestion, food is changed into a simpler, soluble form and is then absorbed from the walls of the intestine into the bloodstream. Only simple substances, with small molecules, can get into the blood.

This breakdown of the food is brought about by the action of enzymes and bacteria. Starches and sugars are changed by enzymes into glucose, proteins are broken down into their constituent amino-acids. Fats become fatty acids, which can either be burnt up as energy or deposited as fat. The minerals are released as salts.

A great deal of the food in the bloodstream is first stored in the liver, to be released when needed. The kidneys filter out anything excess or harmful, which is then passed out through the urine.

Peristalsis is the name given to the muscular movements which impel the food along the digestive tract. When there is something unsuitable in the intestine, or bacteria have irritated it, either the movement is reversed, which causes the animal to be sick, or it is speeded up, which causes diarrhoea.

Nutritionists often seem to change their methods of evaluation and the ones I learned when young are now out of date. They now compare the 'total digestible nutrients' of one food with another. A ready-mixed meal will have been worked out for you! The formula is :-

TDN = Digestible carbohydrates (including digestible fibre) + digestible protein + (digestible oils x 2.25)

The TDN of various foods has been worked out and the idea is that it should help when comparing the cost of foods. Instead of looking at their relative cost per tonne you should be comparing the cost per unit of food.

If you want to get the most out of your pigs, whatever age they are, some of their ration will have to be cereal. Barley meal is the usual feed for pigs, ground and not rolled. Oats are rather too fibrous but they will do if they are finely ground. Wheat is suitable, but if it is too fine it will knead into dough in the stomach of the pig and cause indigestion.

The Lehmann System was introduced into this country just before the war from Germany. It simplifies the problems of feeding pigs. The idea is to feed your fattening pigs up to 50 lb (say nine weeks) on meal. After this start, on creep according to today's methods, the meal allowance (creep, gradually being replaced by meal) is kept at 2½ lbs a day. It is not increased as the pig grows. The diet is supplemented with bulky food as it likes; as it gets older and bigger it will eat more. This is a simple diet that the pig will adjust for itself and the real ingenuity lies in finding sources of food for the bulky part of the ration.

Barley Meal Equivalents. The following food quantities are roughly equivalent to 1lb of barley meal in feeding value.

Wheat Meal 1lb
Ground Oats 1.1lb
Crushed Oats 1.45lb
Potatoes (cooked) 4lb
Dried Potato Slices 1lb
Potato Peelings (steamed) 5lb
Sugar Beet (grated) 3.5lb
Molassed Beet Pulp 5lb
Fodder Beet 5lb
Whey 1¼ galls
Swill (good) 3lb
Grass 10lb
Kale 7.5lb

Swill is naturally very variable but steamed, processed swill, like Tottenham pudding, is referred to above.

Quantities of barley meal (or its equivalent) to be fed to pigs according to live weight.

Live weight (lbs)		lbs barley meal or equivalent per day
20	ad lib creep feed	
30	ad lib creep feed	
40	ad lib creep feed	
50		2.5
60		2.75
70		3.0
80		3.25
90		3.5
100		3.75
110		4.0
120		4.25
Fatteners usually level off here these days		
130		4.5
140		4.75
160		5.0
Even traditionalists level off here		

Remember that pigs vary and always feed according to condition. I will repeat that the amounts given above are to be fed *per day* —usually half at each feed.

How much use is grass to a pig? Quite a lot, but the problem is the variation in grass at different times of the year. The digestibility of the energy in grass can change from 70% for young grass, at the 6 leaf stage, to 30% for the mature grass of late summer. You can say that a big pig will eat up to 14lb of grass a day. About 10lb of grass will replace 1 lb of meal; in actual food value this figure should be less- say about 7lb of grass. But the pig uses up energy wandering about in search of the grass; and also, digesting it takes more energy than digesting meal.

There are things that you can grow for pigs. Fodderbeet, so popular in the 1950's, may well come back as a home-grown substitute for imported grain. In terms of production per acre it does better than barley and the roots have 20% dry

matter, which is quite good for this kind of food. The high water content of all roots is the drawback for pig feeding. They cannot deal with the large quantities needed for proper rationing. But on the Lehmann System roots are useful. Carrots and potatoes can be used, or artichokes. And another plant which can be grown is comfrey, a succulent leafy grower much favoured by many who have tried it.

How much feed should the pig be allowed? If you are using meal, give a sow 5-6lb per day during pregnancy, but *always* feed according to condition. If she is thin, give her more, and cut down the food of a fat sow. Try to work out on this basis how much meal the other foods will replace, and feed accordingly. Roughly 3lb of swill equals 1lb of meal, a bit less with concentrated swill; or 4lb potatoes.

Milking sows need 2lb meal for the sow and 1lb each for the piglets; so a sow with ten piglets, gets 12lb per day or 6lb at each meal. This is arrived at gradually by stepping up her ration over a week or so; don't give it all at once as soon as she farrows.

Keep a check on your scoop. Weigh the meal periodically to check that you are feeding the right amount.

Swill. I have mentioned swill all along in our feeding programmes because this is the ideal pig food. In spite of the present-day difficulties you may decide to get a boiler and a licence- perhaps this could be a joint venture between several backyarders. If you do, there are several points to consider.

First of all, collect the swill in clean containers; plastic dustbins are useful for this. Provide containers with a cover and collect them regularly, leaving a clean sterile one in place of the full one. Always do this, no matter how small the contribution. Be reliable, and in turn ask for a high standard from your swill producers because you want no foreign objects in the food. It is obvious that pigs will not want broken glass or razor blades in their food, but some people throw rubbish anywhere and in a big establishment a swill bin can look like a dustbin; so write something appropriate on the side- in Chinese if necessary! Other things not wanted are too much salt or soda, rhubarb leaves, citrus peel and anything fermenting. Remember the orange peel- it looks harmless, but is not good for pigs. In some places the swill collector is thanked for performing a service, in others he is supposed to be receiving a favour.

Actually pigs are careful about what they eat, as I said before, but do beware of things like small, half-cooked potatoes. We nearly lost a pig that choked on one of these.

Water. Clean water is very important; pigs should have as much as they want except when a sow is newly weaned. Then the water is cut down for a few days to discourage the flow of milk. Pigs bathe in water troughs so they have to be cleaned regularly, and a heavy trough will be kicked about like a football if it is not fastened down securely. The only way to secure a trough is to concrete it in, or wire it firmly to the side of the pen. If you want the little pigs to drink from the sow's trough, remember to make it low enough. I find that an in-pig sow will drink 2 to 3 gallons a day, and double that amount after she farrows and is producing milk.

Wild Food for Pigs

City dwellers can usually find free food for pigs which is surplus to the requirements of man. In the country at most times of the year there is free pig food provided by nature, waiting to be collected. Of course, there are hazards; beware of trespassing – it will be best to get permission to go on private land. On waste ground and roadside verges there is a lot of useful material, but in this case beware of petrol fumes and their lead content, which cover the sides of busy roads.

In Britain it is now illegal to dig up wild plants, but the law still allows the gathering of wild fuits, flowers and foliage so long as it is for personal use and not for profit. Splitting hairs we could say that pigs for sale should not be given wild foods, but pigs for home consumption may have them. The spirit of the law is intended to protect the countryside from wholesale plundering, and quite right too. It is up to us to be sensible and not to take too much from any one plant or area.

There are some plants which you will find particularly useful. Some will fill up the pigs and some can be used as medicine. Comfrey, mentioned in the section on growing food, comes into both categories. In some country lanes wild comfrey grows in profusion; I am not sure whether it is the true wild plant or 'escapes' from a long ago field of comfrey. It grows round the Cistercian Abbey of Fountains and they say you can find it near old coaching inns because it was used as a pick me up for coach horses. This plant is very useful for curing digestive upsets and scouring in pigs. It is a perennial, so you will find it in the same place every year. The leaves look rather like big foxgloves, but the flowers are drooping clusters of small bells.

Garlic is an ancient remedy for all kinds of ills and a general antiseptic. It is a harmless dose for expelling worms if fed in fairly large quantities. The commonest kind of wild garlic is probably *Allium ursinum* or Ramsons (Razzums in some parts). Leaves and bulbs are both used. It is quite possible that a lot of garlic in the last stages of fattening a pig might taint the pork, as the plant taints the meat of wild rabbits sometimes.

Chickweed (Stellaria media) is another common weed, usually found on good land, which can be gathered for the pigs. They love it and it can be found all the year round, even in bloom at Christmas. Chickweed is rich in minerals, particularly copper.

The medicinal value of chickweed is considerable, I have used it to cure cases of scour and it can also be used as a healing poultice for wounds such as pigs get when fighting. Fresh chickweed should be applied to the wound and the poultice changed every few hours.

Fat Hen (Chenopodium album) is also called pig weed, indicating that they like it! Another name is Goosefoot from the shape of the leaves. This plant is the annual relative of the perennial herb Good King Henry, now making a comeback as a vegetable. When young, fat hen makes a good feed for pigs. Later in the summer the stems get very tough. It grows in the vicinity of manure heaps and well manured land.

Hogweed (Heracleum sphondylium) is another plant that pigs ought to like! This is the cow parsnip. In Siberia a form of sugar was extracted from the dried stalks of this plant.

Most green vegetation will help to fill up a pig and give it vitamins and minerals and a small part of its protein and energy needs; but remember to avoid the poisonous plants. Many garden plants would be poisonous to pigs but one hopes they will keep out of the garden. Wild ones to avoid include those in the tables.

Woodland Foods

These were traditional for pigs in the past, and the right to graze woodland, pannage, was one of the basic medieval rights. During the first world war some work was done on finding out the feeding value of nuts. It was suggested that

SOME WILD PLANTS TO AVOID

Plant	Description	Remarks
Buttercup e.g. Creeping Buttercup, Ranunculus repens, also Lesser Celandine	Very common weeds, yellow flowers in grassland and a weed in gardens On banks in early Spring.	Ranunculin is poison produced by this family. It is very irritating to the skin
Ragwort Senecio jacobea	Yellow tall raggy flowers in old pastures	Will normally be avoided by pigs if they are grazing outside
Foxglove Digitalis purpurea	Tall plant with pinkish bell-shaped flowers	Digitalis is cumulative poison
Henbane Hyoscyamus niger	Large yellow and purple flowers June—August. White hairs on stems and leaves. Unpleasant smell when bruised	All parts poisonous, but alkaloids most concentrated in seeds
Honeysuckle Lonicera periclymenum	Climbing shrub with pink and cream scented flowers	All parts poisonous but the berries produce worst effects.

Plant	Description	Remarks
Deadly Nightshade Atropa belladonna	Rather rare; large purpose flowers, June—August	All parts poisonous
Hemlock Conium maculatum (Herb bennet)	Umbelliferous. Purple blotches on stem, mousy smell when bruised	Very dangerous
Yew Taxus baccata	Churchyards, old gardens. A dense evergreen tree	All parts poisonous. Animals may eat it in winter if they are hungry

These and other poisonous plants are not to be feared too much if they grow in small amounts where pigs graze outside. Animals will usually avoid them if properly fed. But do not gather these plants to feed to indoor pigs.

they should be fed to pigs of about 100lb live weight or more, as younger animals might have difficulty in digesting them.

Acorns. These can be collected and stored. They should be spread out in a thin layer to dry because the astringency decreases with the moisture. They have a high starch content and should be fed along with some laxative food such as bran. About 1lb of acorns a day would be a safe amount to feed, going up to 2lb for a large pig such as a sow. This could make up a third of the total ration and it would be a worthwhile saving of meal.

Horse Chestnuts. These are even better than acorns as a feeding stuff, but they have to be dried, taken out of the husk and ground. If this can be done, chestnut meal is the result, and 1lb chestnut meal is equal to 1lb 1oz of barley meal. But unfortunately it tastes bitter and animals will tend not to eat it unless it can be mixed with something more palatable. About a pound per head per day is a reasonable ration.

Beechmast. There is a lot of fibre in the husks and the kernels are small, but pigs enjoy rooting about under beech trees so who can say this food is not for them?

A good guide for feeding adult pigs is to give them what they will clear up in twenty minutes. Never leave stale food in front of the pigs and do not be afraid to miss out a meal if they start to leave food. Some people make a practice of feeding only once on Sundays - for the good of the pigs, they say!

If a watch is kept on the dung, the diet can be adjusted in time; loose dung means sometimes that the pig is getting too much food. If the dung is hard, feed bran or some green food.

7 growing food for pigs

Since the cottager's pig was more or less the dustbin and was fed largely on what nobody else wanted, the cottager might have been startled to think of growing food especially for pigs. But food costs are now so high that the odd corner of garden or field can be used to great advantage if it can be made to produce crops that pigs can eat.

Vegetables are too bulky to provide all the food for fattening pigs, but they can contribute a great deal to a healthy ration. Those normally grown in the garden will all be acceptable to pigs and the gardening books take care of advice on how to grow them; see the writings of Lawrence D. Hills on vegetable growing.

The so-called farm crops are left for us to consider, although I am only thinking in backyard terms and of a small patch at a time. But those with an acre or two might like to grow some of these crops to benefit all the animals on the holding; so I will refer to seeding rates and expected yields per acre. You can scale them down or up to fit your requirements. One difficulty with them is that perhaps the seed for farm crops may be difficult to buy in small quantities, but neighbours could perhaps combine to buy some, or the local feed merchant who usually sells farm seeds might be persuaded to do a small order.

The best thing about pigs is that they will actively help

you to grow food for them. Suppose for a moment that we have a piece of worn-out grassland, mostly couch grass and low yielding weed grasses. The soil is reasonably deep, but it may be short of humus. If it would only grow good grass it could provide part of the pigs' food and grazing for other animals as well.

We take it in hand in early spring, or rather we allow the pigs to do so. First of all it is fenced carefully with two strands of electric wire, and the posts which hold them are stout affairs with extra support at the corners. Pigs like to scratch on posts and can soon knock them over. Inside the compound we erect a simple straw bale hut. Then the pigs come in, without rings in their noses, and eat off the grass. While this is going on they may be given a little less meal, but do not overestimate the food value of old grass.

Soon there will be only tussocks of couchgrass left. Then these will come out, including the roots, and the earth will be turned over by the joyful pigs until you have a fine but very uneven tilth. All the vegetation will be gone, which is why I suggest putting in the pigs in the spring. If they did the job in the autumn, the ground would be left bare all winter and plant nutrients could be washed away.

When the pigs are removed the land has to be evened up again; harrows will do it, or a rotovator. There will be pig muck in the soil and this will make a contribution to feeding the next crop.

At this stage you could reseed the field with a good grass seed mixture and have grazing again within a few weeks. In some cases this will be the best thing to do. But just now we will go through the stages of growing another crop for the pigs.

Potatoes are a good crop to plant at this time. The large amount of manure or compost they require will be good for the land, and any weedy grasses will be weeded out; and the yield of potatoes could be good in a decent year. Potatoes are very good food for pigs as they do not contain the fibrous bulk of other vegetables. We can expect a yield of about 6 tons per acre, so half an acre could produce 3 tons, enough for the household to be supplied with the large ones

while the small potatoes can be boiled for the pigs.

By early April we should be ready for planting, but the very earliest potatoes should be given a miss; their place is in the garden. Potato varieties are divided into first early, second early and maincrop and it is these last which we will grow for the pigs and to see us through the winter. It is interesting to experiment with potato varieties, so make a note of what you plant and assess it for yield and flavour.

The potato 'seeds' are small tubers and they will grow more quickly if they have been allowed to put out sprouts before planting. They are left in a warm place on trays in daylight for a few weeks before the planting date. The sprouts they put out should be firm and green and 1/2 to 1 inch long. In the dark they would develop weak white sprouts. For an acre of land about 18cwt of seed is needed, but the seeding rate is according to variety.

Trenches are dug in the time honoured way, and filled with muck or compost, potatoes need a lot of rich plant food. The traditional day to plant potatoes was Good Friday, because it was the first spring holiday.

The tubers are planted at about 10 inch intervals and then the trenches are covered with soil. When the potato plants start to show through, they should have the soil drawn towards them with a hoe, to protect them from late frosts. Later the soil is heaped round them in ridges so that only the leaves show; it is important, whether the potatoes are for man or pig, to keep the sun off the tubers. If they are exposed to the light, they will go green and will contain solanin.

The weeds will be kept down by hoeing between the ridges. This will improve the crop and subsequently the land. Chemical weed control has now replaced hoeing on many commercial farms, but we can manage without it.

You can start eating new potatoes about July with a maincrop variety, but the bulk of the crop will be ready in September. They should be taken up in a dry spell of weather if possible, so that they can be stored dry. Small quantities can be stored in bags in a frost-free shed, but a clamp is probably the best way to store potatoes.

However you try to pick them all, some potatoes are

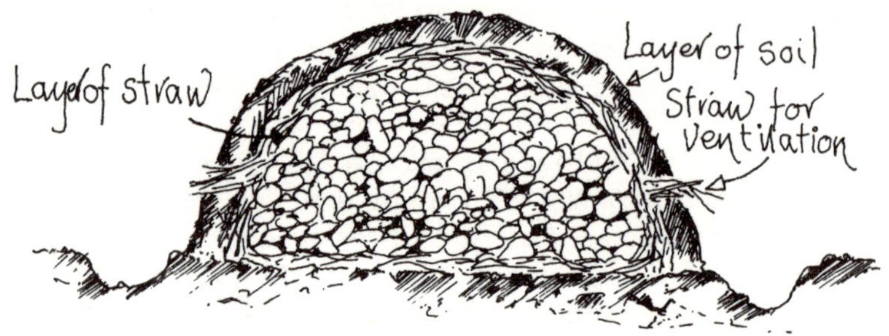

always left. The pigs can be brought back again for a final gleaning and then, if it is not too late, the ground can be sown with grass seed.

A good mixture of grass seeds will be better than a single species. Clover should be included because it adds nitrogen to the soil, through the action of nitrogen fixing bacteria in the nodules on its roots. This benefits the grasses, as nitrogen encourages the growth of green leaves. The clover might not do so well the first autumn, because it prefers warm conditions, but unless it is killed off by early frosts it will grow when the ground warms up in spring.

Perennial ryegrass will be essential in the mixture for general purposes if the plot is to stay down to grass for a while. Its annual relation, Italian ryegrass, should be included for a good yield of grass in the first year. Cocksfoot, meadow fescue and timothy are the other main agricultural grasses useful in various situations and for various purposes. Your seed merchant will be able to advise you on the best seed mixture for your land. Where 'poaching' is likely in winter, there are now tough creeping varieties of ryegrass available under several commercial names. One is called Stadion and I am assured it has been used for the turf of a famous football ground, so it should stand up to trampling by the pigs.

The seeding rate per acre varies; the average is perhaps about 20lbs to the acre, of a mixture of five or six grass species.

Many farmers also include a few herbs in the mixture;

yarrow, ribgrass and chicory are some of the popular ones. Herbs give variety to the sward, which any animal will appreciate. They also provide extra minerals and their deep roots penetrate the subsoil and help with drainage.

When the grass seeds have germinated and there is a green cover, perhaps by spring, the patch can be grazed lightly for a day or two and then left to recover. Light grazing encourages the growth of new side shoots. But do not spoil the new grass by letting it be grazed right down and trampled too much, and try not to cut it for hay in the first year. Of course any pigs that graze it must have rings firmly in their noses.

The grass crop is a very important one, and the way in which the pigs can help shows how they fit in with other animals. In fact the new grass might be too good for the pigs! if there is more to reclaim they can move on.

Kale

This is a farm crop which gives a green bite in the winter to all the animals; it is very good for pigs. It can be fed chopped up and 7 or 8 lbs of kale are equal to 1lb of meal. There are several kinds of kale, but in general it is very hardy and disease resistant and therefore easy to grow.

Marrowstem kale grows to a height of 5ft on good land, but the stems are tough. It is not frost hardy and should be eaten by the New Year.

Thousand head kale is hardier and grown for use in late winter.

Hungry Gap is actually a rape, but similar to kale and the yield is lower.

By using a succession of different varieties it is possible to have a supply of kale from August until May.

Kale is usually grown in rows about 2ft apart, the seed

sown at a rate of 2 to 4lb per acre; the lower rate for thousand head, the higher for marrowstem. If you are only sowing a small patch and prefer to broadcast the seed, twice as much will be needed. The advantage of growing it in rows is that you can hoe it if the weeds are too vigorous. It is not singled. Kale needs plenty of manure, up to 20 tons per acre.

The yield of kale will be about 10 to 20 tons per acre. For pigs it will be best to cut the crop and carry it to them so that it can be chopped up. It can be grazed with an electric fence , and this method of feeding would be possible with pigs if they were on dry, well drained land.

Comfrey

The story of comfrey is fascinating, as told by Lawrence Hills in his book 'Comfrey, Past, Present and Future'. He calls it the gardener's herb because of its many garden uses, but it could be called the backyarder's mainstay too. It is a good feed for pigs.

In some parts of the world comfrey grows wild. This is the herbalists' comfrey, *Symphytum officinale.* The cultivated sort is a semi sterile hybrid, which rarely sets seed. It is a cross between the wild plant and Prickly comfrey.

Strains of the hybrid, which is known as Quaker comfrey in USA, are produced in Britain by the Henry Doubleday Research Association at Bocking near Braintree. Thus the agricultural strain is called Bocking Mixture. The usual sort to grow in gardens is Bocking no. 14 which contains more potassium.

Since there is little comfrey seed about, the plants are started off from root cuttings, which can be obtained from HDRA . They can be planted out at any time of the year except December and January in Britain.

The plant is a perennial; it will be there in the same place year after year, and eventually you will divide it to make more plants. So be sure to choose the right place for it, in full sun and undisturbed. When it is established it can be cut with shears several times during the season — it does not grow in winter.

Comfrey is the highest yielder of all in terms of weight; yields of 50 or 60 tons per acre are not uncommon. So it needs plenty of manure to keep it going. Poultry manure is considered to be the best, but I have found that pig muck grows comfrey very well. This is another case where the pigs can help grow their own food.

The protein content of comfrey is very high for a vegetable and it can provide up to one third of a pig's diet in summer. It is easy to grow and care for and I think you will find it a useful plant.

It is interesting to glance at some of the other uses of comfrey because once you have a supply, you may want to try it in other ways. Here are a few of them:

1. As a vegetable. It is cooked like spinach, minus the tough leaf midribs. Health enthusiasts put the leaves in a liquidiser with water and end up with a herb drink. Comfrey supplies vitamin B12 which is usually only found in animal products.

2. Comfrey tea is an old country favourite. The leaves are collected in summer, dried in the sun and rubbed into fragments for storage. It is usually mixed with Indian tea to make it more palatable, but they say that the mixture is a pleasant smooth drink, very good for arthritis. Country people collect the wild variety for making tea.

3. The allantoin in comfrey is what makes it really popular as a herbal remedy. This substance makes wounds heal and also kills bacteria and has been quoted as a cure for all sorts of ills of animals and man, from athlete's foot to cancer. There is also a sticky substance in the leaves which sets hard when they are pulped and then dried.

4. The high potash content makes it a useful garden manure, particularly for potatoes and tomatoes which need a high potash diet. The leaves are wilted and then laid on the ground or in the trench. It can also be made into a liquid manure by steeping the leaves in cold water for a few weeks.

Fodder Beet

This is another good food for pigs; 5lb beet equals 1lb meal, but it is better to feed fodder beet to older pigs, of over 80lb in weight.

This is so high yielding a crop that it is surprising it has not become more popular for feeding animals; it is not a table vegetable. The high cost of labour has of course, made root growing unpopular on many farms. The chief work apart from sowing and harvesting is keeping down the weeds in early summer, so don't sow too large a patch or it may get the better of you.

There are two main types of fodder beet, the one with the higher dry matter content being normally grown for pigs. At about 20% dry matter it is good feeding value, exceptional for a root. The average yield is about 18 tons per acre. The seed is sown in April or May according to the area, at a seeding rate of 6 to 10lb per acre.

Once the plants are through and showing two proper leaves, they are singled, ie spaced out to about 10 inches apart and the superfluous ones knocked out with a hoe.

The beet are lifted about November, they can be dug out with a fork. The tops, like sugar beet tops, can be fed to stock and are a valuable green feed in late autumn.

You may get as high a yield from the tops as from the roots. In feeding value beet tops are about the same as kale, that is about 7lb equals 1lb of meal. They should be wilted for about a week before feeding to get rid of the oxalic acid, which is poisonous.

Beans

On the lookout for high quality pig foods, it may occur to you that beans are a very good idea. Soya beans can be grown in some countries but the yield is low for a backyard enterprise; in any case, they will not grow in climates such as that of Britain. But the beans grown in cool temperate climates are sufficiently high in protein to give a balanced ration without any additions. 2cwt of beans equals in feeding value, 1cwt soya beans plus 1cwt barley meal. One drawback is that the beans have a lower methionine content ie the protein is not of such good quality, and it is thought that too much bean meal is not good for fattening pigs; it tends to make the meat hard, and the bacon will not be of the best. So beans should only form part of the ration.

Beans can be harvested dry and kept for winter use, but they should not be fed until after Christmas as they may otherwise cause indigestion. The old farmers used to keep them about a year because they thought that old beans made a better food.

Winter sown varieties of beans are mainly grown in the southern half of Britain and in similar climates. One variety of field bean is Maris Beaver. They give a higher yield than spring beans.

Beans sown in spring are of two kinds – tic beans and the bigger horse beans, usually grown for animal feed. Horse beans are sown in early March at the rate of 2.5cwt per acre. The width between rows seems not to be critical. They do not like an acid soil; beans can make their own nitrogen like all legumes, and will leave a residue in the soil – an important point for organic farming; their main need is potash. Beans seem to do better on heavy land than on light soil.

They are now harvested on farms by combine, but on a small scale they can be scythed and bound into sheaves like corn, and then stooked.

They are ready when the pods are black, but it helps if they are left in stooks for about a week to dry out. When they are carried, it should be on a very dull day or very early in the morning; in other words, in damp conditions. Otherwise the pods will split open and beans will be lost. The bean sheaves can be stacked in a rick in the yard and threshed when wanted. The pigs can be turned onto the field for a bean feast of what has been lost on the ground.

The yield is on average about 18cwt per acre; not high, which explains why they are not too common in backyard plots. Aphids are the worst pest.

8 breeding your own pigs

The question of producing your own piglets can be put on one side for a while if you are a complete beginner to pig keeping. It is simpler to buy weaners and fatten them and not to worry about the hazards of keeping a sow. But, of course, if you like pigs, you may want a permanent pig, and if you can find outlets for two litters a year, or perhaps share with a friend, it would pay to keep a sow.

A litter of pigs takes nearly four months from conception to birth; three months, three weeks and three days. Allow two months for suckling the litter and getting in pig again, or perhaps a little longer with eight-week weaning. After weaning the sow will take a week or two, or at most three, to get in pig again. So this means that there are two litters a year. They will be produced for the cost of keeping the sow, plus a bag of creep feed. A sow will eat about 1½ tons of meal or its equivalent in a year. The weaners might be worth say, £320, and the meal would cost perhaps £200. The food could cost a lot less if you were enterprising. Of course, she could present you with ten pigs in each litter for the same outlay, in which case the profit would go up a good deal. The number of pigs in a litter is very variable and it does in fact make the difference between profit and loss. Herein lies the gamble of pig keeping, quite apart from the external effects of fluctuating markets. Good management goes a long way towards getting large litters,

so your efforts are worth something and the gamble is not purely the toss of a coin; but it can feel like that sometimes.

There are points for and against keeping a sow of your own. William Cobbett, in his *Cottage Economy* published in the early nineteenth century, does his best to discourage back-yarders from keeping sows and paints a pathetic picture of piglets going to skin and bone when their mother is taken away. Obviously there was no creep feed in those days; if you wean at the right time neither mother nor young ones mind at all.

One real problem is that of the boar. One or two sows do

not justify keeping a boar; so you have to decide when your pig is in season and then take it to a neighbour's boar. Finding one may be difficult if you do not live in a backyarding area. And also, some pig farmers are so afraid of disease that they will not allow strange pigs onto their premises.

The solution may be to fit yourself up with a really good trailer to go behind the car; or find a reliable haulier of

The Common Boar

small loads. Then distance will not be so important. The days when you could walk a sow down the road to the boar are more or less over! In any case, one of the habits of a sow in season is to stand perfectly still, which is no help if you are walking her anywhere. The other signs of heat are a deep grunt and reddening of the vulva.

When ready to travel you can seek out - by advertising if necessary - some enthusiast who is willing to allow you to use his boar and who has a good boar of the breed you want. Boar licensing has been abandoned, which means that anybody can keep one for breeding; but I should hardly think that this has led to a rapid decline in standards. Breeders will keep the best boar they can, in their own interests.

Even more useful would be a boar owner willing to board your sow for a few days over the time when she is in season. It is quite possible to arrive at the boar to find that she is not ready after all. This is because some sows put out stronger signals than others and it is only the boar who will know for certain whether she is ready or not. If not, it is rather irritating to have to take her home again, although this is fairly usual, even for experienced breeders. A day or two on the boar's premises will ensure that she is served at the correct time.

In fact, all along the line from the mating onwards you will need more management skill with a sow than is needed to rear porkers. The farrowing itself may keep you out of bed and you will have to see the litter through its early stages. If you want to go away for a weekend it would be easier to get a neighbour to feed a pork pig than to look after a sow and its litter.

Having said all this, there is no doubt in my mind that keeping a sow is so interesting that I would not like to buy weaners, in spite of the attendant problems. To justify keeping a sow you can consider that your weaners will be cheaper than you could buy them, and then the surplus can be sold to cover the cost of the ones you keep. Born on the premises, they will start life with built-in immunity to the germs around them and they should therefore grow on well without the check in growth which nearly always occurs when pigs are brought to a new home. There is great satisfaction in rearing a batch of healthy piglets; if you choose a rare breed it could be even more satisfying as you may be helping to save it from extinction.

Let us assume, then, that you have bought a gilt in-pig, and follow her fortunes through six months or one breeding cycle. We will call her a gilt so as to take in any possible complications. A sow should be easier as she will know what to expect.

First of all, her accomodation should be prepared. For farrowing indoors she will need the pen we saw earlier of about 100 square feet. The walls, whatever they are made of, need to be smooth so that the pen can be thoroughly cleaned and disinfected between litters.

To clean a pig pen properly it needs to be done with a pressure hose or steam cleaner. This is to get it really sterile. There is no need to go to such lengths with backyard pigs, thank goodness; they are hardy enough to stand a few germs. But perhaps we should aim for a compromise, and make an effort to reduce the number of bacteria by removing the visible dirt. It will look better to have new pigs in clean surroundings and it may help them to survive their first few days.

The easiest way to clean the pen is to soak it with cold water as soon as it becomes vacant. Keep soaking it until you wash it; do this with hot water and washing soda, which is cheap and effective. A scrubbing brush with a short handle, called a churn brush by country ironmongers, is the best tool to use. Get the dirt off walls and floor, leave it to soak again and finally rinse. Try to leave the clean pen until it is quite dry before the next pig goes in; the longer it is left, the better.

The basic pen will need some additional 'furniture' for housing a sow and litter. A 'creep rail' which can be tubular metal or timber, right round the pen about ten inches from the floor and nine or ten inches from the wall will give the babies a space for escape if the sow flops down suddenly.

At the back of the pen is the 'creep area', about 3ft wide. Rails keep the sow out of here, and must be fairly strong. Usually there are three rails at ten inch intervals from the ground and with a central post for strength; the little pigs eat in this area and lie under the lamp away from the dangerous bulk of the sow.

Some modern farrowing pens have moveable partitions to keep the sow confined for a few days after farrowing; or they have an independent farrowing crate. This may interest you if you lose a lot of babies through crushing, but you will be able to manage without it. The creep area is normally enough protection if it is well arranged.

To be attractive, the creep should be warm and free from draughts. In the winter a lid can be made from hardboard to keep the heat in. The pigs will need an infra-red lamp; the ones with screw holders are used for pigs, rather than the household bayonet fixing ones. The lamp is a very necessary piece of equipment if you want your sow to raise a large litter. It warms the whole pen, keeps the piglets growing and attracts them away from the sow when they are not actually drinking. The sow is happy to lie and watch them basking under the lamp - in fact, I have seen sows pushing their offspring gently under the lamp in cold weather.

Once the quarters are ready, the pig should be brought into them in plenty of time so that she will have settled down before the litter is born. Pigs need time to adjust to any changes and it helps to do things gradually.

You will know what day she is due to farrow, but be pre pared a day or two beforehand. Gilts and old sows often farrow a day or so early, while sows in their prime are frequently a day or two late.

For the last week of the countdown, give her rather less food.

I have found this to be very helpful; it gets her into a fit state for easy confinement. A bran mash or two may be a good idea. If she has been running outside, let her go out during the day and put her in the farrowing pen at night. If you spend some time watching the pig you will know how she normally behaves, and anything unusual you will spot straight away. There are several signs by which you can tell that labour is approaching.

A few hours before farrowing the sow or gilt will start to get restless. She may have milk in the teats. She will run about with straw in her mouth, making a bed and will possibly try to demolish her pen. She may develop a bad temper and chase you away. At this stage I like to switch on the lamp, to get the place warmed up for the piglets. Good-tempered pigs often get very annoyed when I go into the pen at this point in their confinement, but later on, they either ignore me or are quite glad to see me.

This restless stage may last for a few hours, but the next stage will come and the pig will lie down in her nest. Gradually she may go into a coma and start to give birth, but sometimes the first piglets will be born while she is still in the restless state. This is when it pays to be present - you can whip them away and put them in a box under the lamp to dry off. When the mother has settled down, they can go back to her for their first feed.

This often happens when a gilt farrows and it frightens her. She may panic at the sight of the first piglet and kill it, as I said before. This is a time when drugs are useful; when a gilt is really upset a tranquilliser is the best thing to use. The old-timers used to give the pig stout or beer, but a really frantic pig is not likely to drink. We always have a bottle of either *Stresnil* or *Suicalm* and the upset pig never seems to feel the needle go in. This drug makes the pig relax; you are supposed to leave her quite alone for about twenty minutes so that she can feel the full effect. If you stay with her, she may fight against the drug and it will then not be so effective. Gradually she will lie down. The piglets will continue to arrive and she will lack the energy to snap at them. Soon you will hear the proper maternal grunts as she gives them their first feed. This treatment has always worked, and since we started using the drugs, we have never found any difficulty in getting a gilt to take to her litter. The dosage needed is very small. 2cc will do the trick sometimes but with a big pig or a very hysterical one you will need the full dose of about 6cc. Instructions are given with the drugs, which are available from the vet or from farm chemists.

When a sow feeds her young she gives a certain grunt; a gilt that makes this noise has got the idea and in time loses her fear. The drug carries her over the first hurdle of getting to know the new arrivals, and it works in the same way for mixing batches of pigs. Over-dosing means that you knock them out completely, and if you do this it is not so effective; because to work properly the drug must allow the pigs to get to know each other, while feeling too sleepy and relaxed to fight or be afraid. If they are all snoring in a pile, they will not learn about each other, will wake up as strangers and may start to fight; but if their smells have become mixed by then, perhaps the fighting will not last long. We have strayed a little from the point, but these drugs can be useful so I thought them worth a mention; they are also used to calm excitable animals on journeys.

If the sow is restless during farrowing she can easily trample on the first few piglets before they have struggled to their feet. In this case, I pop them into a box and put it under the lamp. If this upsets the sow, one can be left with her. I used to remove the piglets at birth as a regular procedure because they are a little helpless when wet. After a few minutes they dry out and can take care of themselves. But now I leave them with the mother unless they really seem to be in danger. When they stay with her, she does get the job over more quickly because of the stimulation that their sucking produces. It seems best to leave the sow alone if everything is going well; but keep having a look at her to make sure that it is.

If the sow has a few piglets and then no more, but still seems to be trying, she may be in difficulties. Over an hour of trying with no result probably means that she will need help, but it's rather hard to say how long you should leave her because some are naturally quicker than others to farrow. If you are a complete beginner, it may be wise to call in the vet; but with experience you can assist the sow yourself. The procedure is to wash and disinfect your hands and arm, soap your arm or grease it and gently feel inside. With practice you will get to know your way about; there may be a piglet stuck in the passage which she cannot expel without help. But don't pull until you are sure he is ready - there may be a foot doubled back which needs straightening. It is very satisfying to bring out a piglet that would have died - they do die if they have to wait too long - and also to relieve the sow, but on the other hand there is a risk of introducing infection when you make an internal examination so it is not a thing to do unless absolutely necessary.

It is always hard to decide when a sow has finished farrowing. Towards the end she will expel the placenta, but there may be an odd one after this. At any stage there could be a piglet born encased in membrane which will need to be freed if it is not to suffocate; that is why you can save piglets by staying with the sow, and make your contribution to larger litters. At any stage too, a pig can be born dead. In a big litter this does not matter too much but it's disappointing to blow into a piglet's mouth and find no flicker of life; it is always worth a try because seemingly lifeless pigs can be brought round, but dead pigs do happen and unless you seem to have a great many, it's nothing to worry about.

Once you have removed the afterbirth and tidied up, the litter can be left alone with the mother for a while. It is always a good idea to put the babies under the lamp for a few minutes to show them where it is. At first they tend to stick with the sow, but gradually the warmth and the light attracts them. With a large litter, try to watch them feeding to make sure that they all have one teat. Very soon they settle down to one particular teat for each pig, which is why an extra piglet does so badly.

Some people clip the eye teeth of baby pigs to avoid their biting the sow. If she has a large litter and their teeth are very sharp this might help her to settle down; sometimes she will lie on her stomach and refuse to feed the babies and if she has plenty of milk, their teeth could be the cause. Teeth clippers can be bought from ironmongers in the country. Altogether there are eight teeth to be clipped, two at the top and two at the bottom at each side. I find it difficult to hold open the strong little jaws with one hand in order to clip the teeth with the other, but if someone helps by holding the piglet and opening its mouth, it becomes a simple enough job. I only do it when the sow shows signs of distress; perhaps it will be one litter in ten or more.

Outdoor pigs should not need iron, but pigs inside should be dosed at three to four days since the sow's milk is deficient in iron. There is a paste which is put onto their tongues with a spatula - they do their best to spit it out of course. Or there is an iron injection, 2cc per piglet in the ham with a very small needle. Haemalift is one of the brand names.

At the three day stage the piglets' tails should be docked if this is going to be done. It should be possible for you to avoid this by not selling weaners to farmers who object to pigs with tails. The theory is that fatteners will bite each other's tails and cause much suffering, as well as economic loss. No doubt this is so under certain conditions, but I have never seen an outbreak of tail-biting myself. If there is no way of avoiding the operation, get it over with while the pigs are very young; cut off the tails within about an inch of the rump and use surgical scissors. Dip the stump in iodine.

If you are thinking that there is not much left of the piglet after this, what about castration? Gilts and boars can be kept together up to about six months.If you intend to sell weaners the males will have to be castrated unless you can find a buyer who agrees that it is not necessary.Unless you know a real expert at this job, get a vet to do it for the first few times and watch him closely. Farmers' supply stores sell scapel handles and blades, so in time you can do it yourself. Use a fresh sterile blade for each session. The younger the piglet the better; very small ones are difficult to handle, but even so, some people do the job when they are only a few days old. In any case do see that it is over by the time they reach three weeks of age.

At four days to a week you can give the piglets creep feed. They will not take much at first, so only put out a small amount and see that this and the water is fresh.A creep trough will be needed soon to replace the churn lid or small container used to start them off on creep; and the main thing is to secure that trough right at the back of the creep, well out of the sows reach. At the moment of writing I am sure that a sow of mine is stealing the creep from her babies; she was so intent on this that she got her head stuck under the bottom rail this morning.

If the piglets look thin in spite of all your care, consult the vet. If shortage of milk is the trouble, the sow's milk can be supplemented by a shallow dish of warm milk in the creep. Stragglers in a large litter can be kept going in this way. There are some expensive sow-milk substitutes you can buy and they are extremely good, so if it is a matter of life and death they are worth buying. But for helping out in a mild case of milk shortage, cow or goat milk will do very well.

Let us hope that the piglets do not need extra milk and that they are sleek and fat and very lively. Plenty of straw in the pen is a good thing for piglets as well as for adult pigs as it gives them something to play with, but be careful for the first few days. When piglets are very small too much straw in the pen can hamper their movements and stop them getting out of the sow's way.

Weaning, then, will be the next hurdle. It used to be standard practice to leave the piglets with the sow for eight weeks. If your pigs are running outside, this may still be a good age to wean. But indoor pigs are eating a great deal of creep feed at five weeks, and if they are well grown and independent, this is a good time to wean them. The state of the litter is more important than the date. The modern idea is to wean at three weeks and this I do not agree with; nor do I like the practice of removing piglets from the sow at a few days old and rearing them in cages. The sow is the best rearer of piglets that there is; it is just stupid not to use her for the job.

Weaning is simple enough - take the sow away and give the

piglets lots of straw and plenty of water. If they are outdoors perhaps the piglets will be removed and the sow left, but it is better if possible to leave the little ones in their familiar pen. Well-grown piglets will not mope when the mother leaves, they are more likely to celebrate by rushing round the pen like chil-

dren let out of school. But the sow and newly-weaned litter should be out of sight and sound of each other, or else when feeding time comes round there will be regrets.

The sow should be given less food and very little water for a few days after weaning. Watch her carefully and if she continues to make a lot of milk, give her a dose of Epsom Salts (a small handful in meal and a little water). Mastitis or inflammation of the udder can develop if this is neglected.

When she has lost all her milk the sow will need feeding up again. By now she will be rather thin, but this is normal. In a day or two she will come in season and then you will need a boar.

The problem of a boar was referred to earlier and it is a real one. But there are ways round it. It is possible that a few backyarders could combine to keep a boar. The economical number of sows per boar used to be about 12 to 15, but it will be more now because of the cost of his food. This is the trouble - otherwise boar keeping is not too difficult; they are usually amiable but they need a tough pen, food and enough work to keep them in trim.

Artificial insemination for pigs seems to be a definite possibility. There are difficulties, but in the future these may be ironed out and AI could then be really useful to the small pig keeper. The problems now are the low conception rate, which is improving but is still not as good as the results achieved with cattle; and transport. There are few centres so the semen usually has to come by post or rail, which seems to upset it sometimes. When it arrives you have to inseminate the sow yourself; no doubt this is a technique which can be learned.

If you would like to consider AI, the MLC (Meat and Livestock Commission) runs a pig AI service, complete with travelling inseminators, for parts of Yorkshire and for the Taunton/Honiton area. Other parts of the country are covered by their semen delivery service by railway and post. You must enrol with a centre a week or two before you are likely to need their services. The addresses are:-

MLC Pig Breeding Centre   Leeds Road, Thorpe Willoughby,
Selby, Yorkshire.
West Buckland, Wellington,
Somerset.

9 pig health

I have labelled this chapter *Health* instead of *Disease* because this seems a more positive approach; it must be better to promote health than merely to treat disease. If animals are healthy they will have more resistance to disease.

If your pigs are bright and frisky, with a sleek coat, they are healthy and it is your job to keep them so. There are several ways to do this; modern pigs are healthier than their ancestors of even thirty or forty years ago, partly because nobody can now afford to overfeed them.

Their surroundings of course play a great part in keeping pigs healthy. I mentioned the fact that the farrowing pen should be scrubbed before each litter arrives. Between any batch of pigs and another it is just as well to clean the pen so that numbers of germs are kept down to a reasonable level. Drains, too can be a breeding ground for disease and need to be kept in working order. Good ventilation and absence of draughts are important.

Another source of germs is other people's pigs and it pays to keep them away from your own as much as you can. If you visit a friend's pigs, disinfect your boots before and after the visit. Sows and boars have to meet, of course, this can't be helped. But if you buy any new pigs, keep them apart from your other pigs for a few weeks. Pigs get accustomed to the bacteria of their home surroundings and young ones are born with a certain amount of immunity to the mother's surroundings. Bought-in pigs are without this immunity to your bacteria and they will bring others with them to which your pigs may be susceptible. This is why they should be kept apart.

When it comes to health the outdoor pig has a great advantage. Pigs will do better outside except in extreme conditions; they get more exercise and occupation, and all their requirements in minerals and vitamins. Sunshine, too, does a lot to keep stock healthy. But you may have no choice; perhaps your pigs will have to stay indoors. By giving them fresh turf regularly you can make up for some of what they miss, at least.

You will build up your own experience of how to treat pig ailments when they do occur if you ask intelligent questions when you have to call in the vet. Vets are busy people but they often appreciate a quick cup of coffee and this can be valuable time if you ask the right questions. I have found that once you have taken their advice, vets don't mind if you use your acquired knowledge and treat minor ailments on your own. But in serious cases always get professional help. Pigs can get ill very quickly and they can die before you know where you are. This is a great shock and a loss as well; the vet's fees are small by comparison. Do not be afraid that your enterprise is too small to be bothered with. If you are interested in your animals the vet will be sympathetic; what vets hate is indifference.

There are a few textbooks with descriptions of pig diseases and treatments, so I will not go into these in detail. Instead I will mention one or two things which perhaps you should know about.

Notifiable Diseases. These are diseases which are considered so serious that authorities wish to know at once when an outbreak has occurred. The vet is responsible for making the diagnosis; and if it turns out that one of these diseases is suspected, the police are informed and the MAFF veterinary department takes charge. A 'standstill order' is made, forbidding the movement of livestock in the area, in an effort to prevent the spread of the disease.

Anthrax is probably the most frightening of the notifiable diseases from a human point of view because it can affect man, but cases are usually isolated so there is less fear of an epidemic.

The spores of the anthrax bacillus are very resistant and can live in the soil for years; but in Britain the climate is too cold for Bacillus anthracis to come out of its spore and multiply except when it gets into the body of an animal. Once it does this it is deadly; but with the pig it is apparently not quite so speedy as it is with cattle and sheep, which can be dead in a few hours. Pigs sometimes get a large swelling in the throat as a first symptom.

The main thing to remember is that anthrax may be suspected in any case of sudden, unexplained death among your stock. The vet will take a sample and until he has the result it is wise not to touch the animal, because if there happens to be anthrax, there is the danger and there are regulations to be complied with. The carcase must not be opened and it has to be buried six feet deep in quicklime, or burned where there is not sufficient depth of soil. Cases of anthrax are not common.

Foot and Mouth Disease. and the similar SVD are caused by a virus and the severe measures which are taken when an outbreak occurs indicate its infectious nature. It can be transmitted by anything, even it is thought by birds and the wind. So if an outbreak of foot and mouth occurs, all the cloven-hoofed animals on that farm are slaughtered. This is a frightful thing to happen, but the slaughter policy has kept our stock free of this disease and most people in the end think that it is worthwhile. It is apparently a painful disease, with blisters on feet and mouth of the affected animals. They would perhaps recover if they were not slaughtered, but it is said that they would never be the same again.

Swine Fever is also infectious and notifiable; it is a virus which can sweep through a whole herd of pigs very quickly, but it is difficult to spot because there are no symptoms peculiar to this disease. Animals are off their food, and miserable. If the vet suspects swine fever he will do a post-mortem.

METHOD OF HOLDING SMALL PIGS FOR THE ADMINISTRATION OF MEDICINES.

Swine Erysipelas is *not* notifiable. It is infectious and it occurs with varying severity - there is a mild, an acute and a chronic form. Purple blotches on the skin, often diamond-shaped, are easy to recognise but they do not always appear. Sometimes pigs die but it isn't always fatal. If you have a problem with this disease your pigs can be vaccinated against it.

Parasites. Pigs are not quite so much at the mercy of parasites as are sheep; but there are several which live on pigs and in large numbers can affect their health.

Lice. These are quite common and if you buy a pig you may well find one or two moving about on its back, particularly on the shoulders. There is only one type of pig louse and they have a simple life history. Female lice lay eggs which stick to the hairs on the animal's back; they hatch out in a week or two and in a few days have moulted three times, matured and are ready to lay eggs themselves. There are louse powders on the market which are sprinkled on the back of the animal - Coopers and ICI both make one which are both effective, the Coopers product smells quite pleasant and the ICI one does not. A cheaper way is to anoint the pig with sump oil, which makes your pig smell like an old bus but which discourages the lice. In bad cases two dressings may be needed, the second about a fortnight after the first so that the eggs have hatched since the last dressing but the females have not reached maturity.

Mange. The mange mite is an unpleasant creature. It is a tiny parasite, too small to be seen by the naked eye, which lays its eggs on the host, and these hatch into grubs. The mite injects an irritating fluid into the punctures it makes when feeding on the skin of the animal, which seems an ungrateful thing to do. This causes red pimples to be raised on the skin, which becomes scaly in time, and the hair falls out. Animals with mange are so uncomfortable that they lose condition; it is not so obvious when a pig has mange because you may not notice the rash and the loss of hair at first. But bear it in mind so that if it does occur you will spot it before the animal suffers too much.

Fortunately there are skin dressings which will clear up mange fairly quickly. They have to be scrubbed into the skin energetically but the pigs are so pleased with this process and

MODE OF SECURING LARGE PIGS WHEN GIVING MEDICINE.

the relief is so immediate that it is a pleasure to do it for them. If a pig looks rather pink as though it has been out in the sun this may be the first sign of mange - look closer. The dressing can be used to good effect at any stage, but the less scaly the skin, the easier the job will be. One dressing is called *Alugan*. It is made by Hoecht Pharmaceuticals of Hounslow.

Worms. All species of domestic animals harbour round worms, which are probably the most common parasite, but in general each animal species is infested by a different species of worm - which explains the value of mixed grazing. Pigs can be afflicted with stomach worms and lung worms, both of which will cause unthriftiness. Lungworms only occur in pigs which live outside because they are picked up from the grass, their alternate host (in which they live out part of their life cycle), being the earthworm.

It is easy to dose for worms using pellets (Coopers) which are given with the food. Sows can be dosed routinely about twice a year. If you prefer herbal remedies you will probably know that the herbalists recommend garlic for all kinds of worms.

The Regulations. I have mentioned the regulations about notifiable diseases, but your vet will be on hand to help you to deal with emergencies of that sort, should you ever have them.

There are however a few rules which you will be expected to observe by yourself and we will consider them so that you are prepared.

Movement Register. This records all cloven hoofed animals coming onto and going off the premises. There are books ruled out for the purpose but you can make your own; buy a hard-backed notebook and rule four columns across a double page.

Date	Type of animals	Premises from which moved	Premises to which moved

It is possible that nobody will ask to look at your book for a long time but be assured that one day some official will want to see it; so keep it up to date.

Movement Permits. These vary according to outbreaks of disease. At the time of writing a permit is needed to move pigs other than those going for immediate slaughter. SVD is the problem in this case. Application forms and permits can be obtained from police stations (though not small village ones) or from the county council special officers who deal with this and with the sheep dipping regulations. A permit will only be issued if no pigs have been moved onto the premises during the previous 21 days.

Swill Boiling. The Ministry of Agriculture has taken over the licensing of swill boilers from the police, so you apply to the Animal Health Department of your MAFF Divisional Office for this. They will give you advice and tell you how to arrange your swill boiling to comply with their regulations. Apart from having to boil it for an hour, there are other rules, designed to keep the raw swill away from the cooked swill. Feeders made a great fuss when the regulations were tightened, but the blame must lie with the careless ones. I know people who have never cooked their swill and it is they who have given it a bad name.

Useful to Know:

The normal temperature of a pig is 102 deg. F.

Respiration 20-30 per minute.

Pulse 70-80 per minute.

Heat period, about 2 days, occurs 3-10 days after weaning and recurs every 20-21 days.

10
the harvest

Jambon — Pieds et Tête de Porc
Boudin — Andouilles
Grillade et Rôti de Porc
Épaule de Porc

Punch à la Romaine

Dinde bourrée de foie gras truffé
Écrevisses à la nage
Bombe glacée Dame Blanche
Desserts — Fruits

Xérès sec — St-Julien en carafes
Musigny — Moët & Chandon

So now we come to the crunch, when you will reap the reward for your efforts at pig keeping.

How will you know when the pig is ready for killing? It may be best to kill at the conventional size for a start, until you find out how you like your pork and bacon. Borrow a pig weigh if possible. If you have an orderly mind you can keep records and work out exactly how much it has cost you - this can be very interesting; but the glory of backyarding is that if you do not like recording, you needn't bother. If you have enjoyed keeping pigs, they have had a happy life, the harvest is good to eat and little money has changed hands during the process, you don't need figures to tell you that it's a success.

There is a thing called a weigh band which will give you a fair estimate of a pig's weight. It is sold by Dalton's of Nettlebed, Oxfordshire.

These are the usual pork live weights:

Light Pork - 90-100 lbs.
Medium Pork - 110-149 lbs.
Cutters - 150-190 lbs.
Heavy Pigs - 240-280 lbs.

Light pork is rather wasteful but very quick to produce; I should think it would not have a great deal of flavour.

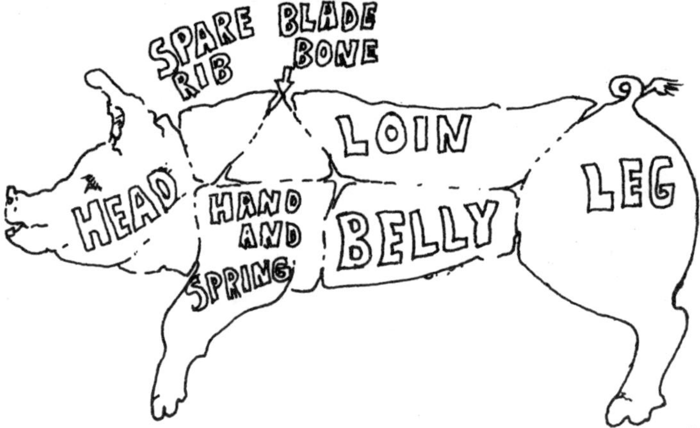

The time of year for the pig killing is still important. A lot of backyarders wait for the first frosts before they think of killing a pig for the winter and this is sound practice. One could think that with a deep freeze waiting the time of the year is not critical, but it is better to have cool weather when you are dealing with a lot of meat, especially on premises not really designed for the job.

As I have said, it is now illegal to kill your pig at home if you intend to sell any of the meat -- although it is still done in remote places on dark nights. If you want to do all the traditional things and get the pig back as soon as it has been killed and bled, in theory you can use the public abbatoir. In practice this sometimes is a closed shop and it may be difficult to get them to do it. This will need investigating before the pig is ready.

It is however, easy enough to get an abbatoir such as FMC to kill and dress the pig, and to joint it too. We paid £3 per pig recently for killing and dressing only. Remember not to feed the pig before it goes away, but give it plenty of water.

Just in case you can establish rapport with the slaughtermen and can get hold of the blood, this is the time-honoured way in which black puddings are made.

Black Puddings. Catch the blood in a bowl and keep stirring with a wooden spoon until it is cold - to prevent it congealing, according to all the old people I have consulted. They are all very particular on this point, although curiously enough I have been advised that stirring will not stop it congealing in any case. I must admit that when I tried, the stuff did stick together. After the blood has cooled there is then no danger of clotting, so perhaps this is the critical thing. The puddings are made after the pig has been cut up and there is fat available.

Method. Stir coarse oatmeal into the blood until it is fairly stiff. Season well with salt, pepper, marjoram and a little thyme. Add small pieces of diced pork fat. Put into a well-greased dripping tin and bake in a moderate oven. Cut into squares when cold and serve fried with bacon and egg; this is of course very nutritious. Some recipes add chopped onions and the herbs vary; suggestions are diced mint, pennyroyal, powdered sage. Rice can be substituted for oatmeal.

Dressing the Pig. If the pig comes home as it was killed, it will save money but there will be the job of scraping and cleaning it waiting for you. The help of a veteran backyarder is invaluable for the first time you try this. The first job is scraping and it should be done as soon as possible. Plenty of hot water will be needed, just off the boil; lay the pig on a concrete or stone floor and pour the water over it, scraping as you go. We use the bottom of an old candlestick to scrape off the bristles a sharp knife will cut the skin. Some people immerse the pig in the water to loosen the bristles.

When scraped and washed clean the pig is hoisted to the roof and disembowelled. A clean old fashioned wash house or dairy is the best place for this operation, but you may have to adapt the garage.

Hoisting up the pig is quite a job. Cut through the back of the hock of both back legs and insert a strong stick between

the tendons and the leg. This is called a gambrel and is supposed to be one of the few Celtic words remaining in English from 'cam bren' or 'bent wood'. So you see how old a tradition it is. This gambrel gives you something to get hold of; the pig is then hoisted up over a beam if possible, on a stout rope, or better still, a pulley.

Slit open the carcase down the middle of the belly; do it by cutting the wall but *not* the intestines - professionals move two fingers down, one on either side of the knife, and pull out the wall in front of the knife. For this job you need a large knife and a sharp one. The rectum is cut out and removed with the intestine, which will probably fall out.

Lungs, windpipe, gullet and heart are next taken out - these are what is referred to as the 'pluck' in old books. The liver is saved and the gall bladder is discarded. The head is severed behind the ears up to the backbone on either side.

After this, you leave the carcase to cool; chill it if possible. A good sharp frost will help you in this, which is why the old timers waited for cooler weather before they started. Your pig is dressed and in this state you will normally get it back from the abattoir. You may feel that it is worth £3 to avoid the jobs of scraping and taking out the innards, but I sometimes think it is as well to know how to do these things.

When you come to deal with the pig, the whole carcase is split right in two with a very sharp, light meat chopper, and of course the head is removed.

Pork. It may be easier to cut up your first pig for the freezer and to forget about bacon until you get more used to handling large amounts of meat; the sheer volume can be rather off putting but be assured you will get used to it.

They say that pork should be put into the freezer within 24-48 hours of slaughter and it certainly isn't long before pork starts to go off a bit in warm weather (which is no doubt the reason for the old taboo on eating pork unless there is an R in the month). In winter it is not so critical; but wait a little for the meat to set before jointing it. As soon as the joints are cut and wrapped they should go straight into the freezer. I think it is important to date the pork you put in, so labels will be useful - you can get them ready beforehand. The books on deep freezing usually say that pork will freeze well for six months, but we have kept it a year with no harmful effects.

A whole pig may take up a lot of room in your family freezer. If you find that you are short of space, there are several space economies that can be made. Joints can be boned so that you are freezing solid meat; bones take up room and they may also pierce holes in the polythene freezer bags. Surplus fat can be trimmed off, which may be a good thing because the bacteria which attack fat can live at low temperatures. Some of the meat can be preserved as bacon (see later)

or can be swapped for other things that you need.

On the other hand, if there is plenty of room in the freezer, you could freeze your ham or bacon after giving it a light cure; this would be playing safe. Although they tell me it would probably not taste the same as proper home cured bacon.

Learning to cut up meat is a process of trial and error when you do it by yourself. There is a book called Meat Technology which tells you how to do it; no doubt any manual for apprentice butchers would be helpful, although these seem to be scarce. I get the feeling that there is a sort of closed shop about butchering and that the old guilds are preserving the 'mystery' as much as they can; but this may be fanciful. The best way to learn is to watch it being done, as many times as possible. When we first bought a freezer we invited a butcher's cutter-up to do the job for us; he seemed to be pleased to be paid in meat, but this might not always be the case. We all watched avidly and my brother, who was to be our butcher, was the nearest. After watching two or three times, my brother had a go himself. There is no doubt that it is an art and it takes some learning; but if you can put up with a little inconvenience when carving at table, you should soon be able to joint your own pork. This is what the art is all about - making life easy for the person who carves the meat. We get rather anonymous joints in our household and sometimes the carver swears; but it doesn't really matter.

It does not take long to grab each joint as it comes off the chopper, pop it in a bag, squeeze the air out, seal it with a twist of soft wire and label it before freezing.

Cutting up the head is the most grisly part of the job, and somebody in the family may protest; but it seems a pity to waste such a lot of good meat, so we will press on with it.

In most of the recipes the head is soaked in brine for a few days, after it has been thoroughly cleaned. Then it is put into a large pot and brought to the boil, in water with a blade of mace, cloves, herbs and white peppercorns. Simmer gently for four hours.

When cold this water will set into a jelly

and you can skin off the fat. The meat is cut off and cubed, as is the tongue. It is set in small bowls with the jelly and turns out like a mould when set. You have made brawn. At its best brawn is delicious, but you need some resolution to make it this way, with the head staring at you from the pot.

For those who dislike boiling heads, there is a rather less cannibalistic way to do it. Cut off the raw meat from the head in small pieces, salt it for a couple of days and then boil it up.

Bacon. There are many different cures in different parts of the country and the variation in home-produced bacon is amazing to anybody who has only sampled the factory product. Once you have learned how to cure bacon properly, you can then experiment and find the variety you like best. There are several recipes in a book called Farmhouse Fare, published by Farmer's Weekly. Curing is simply the process of drying out the meat with salt to preserve it, but each county seems to have a different method. Some of the cures are sweet and use molasses; some are dry and some are wet.

TILE WITH ARMS AND CREST OF THE BACON FAMILY.
South Kensington Museum.

We usually cure a side and a ham. We use the Yorkshire method which is one of the simpler ones. The side is put down on the slab skin uppermost; it is all rubbed well with salt on both sides and left covered with salt. The rubbing is repeated for three days, along with a sprinkle of sugar and saltpetre. The side is left in the salt for about two weeks, then washed and dried and hung up in a cool, dark pantry.

When curing a ham, you have to be extremely careful because it is difficult to get the salt to penetrate right into the middle. And if you do not, the ham goes bad, which is a dreadful waste. Hams are left in salt for longer than sides, say three weeks. With a steel used to sharpen the carving knife you must make a cavity down beside the ham bone, right in the middle of the joint, and pack the hole with salt and sugar and saltpetre (potassium nitrate).

If there is plenty of salt about you will not go far wrong; it must be impossible to use too much. The bacon will of course be saltier than shop bacon, but this can be remedied by soaking it in water for an hour or so before cooking it. A piece of bacon which you may cut off to boil can be soaked overnight. Only in very hot summer weather will there be a danger of the bacon going bad before it is properly cured; so once again, the best time to do it is in winter.

The old folk tell me that a ham should be kept for at least a year to allow it to mature properly. They used to wrap the ham in a muslin bag to keep the flies off it, but never surround it with anything that will stop the air getting to it. This is most important.

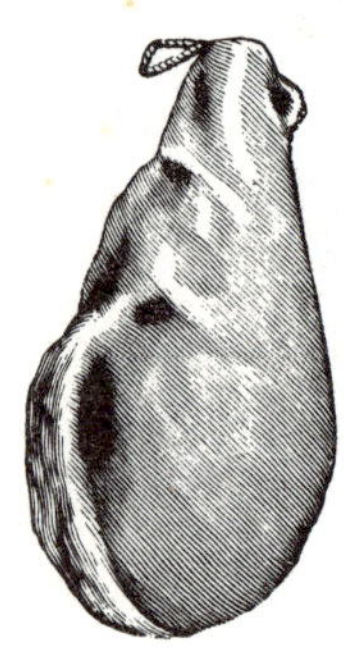

Bacon factories do not go on like this at all. They use the brine cure, and they inject the liquid into the meat. You can home cure bacon in brine - without the injections - by leaving it totally immersed for several weeks, or by pouring the brine over it every day.

A man I know cures his bacon, which is said to be very good, in a polythene dustbin by leaving it in a brine solution for three or four weeks. He gets the ingredients ready mixed in a packet from his local butcher.

If you would like to make your own brine mixture simply add to four gallons of water, 8 lbs of salt, ¼ lb saltpetre and ¾ lb brown sugar. Boil the mixture up together for twenty minutes and let it cool.

Saltpetre tends to make the bacon hard if you use too much, but it seems to be necessary since it is included in almost all the recipes. It is saltpetre which gives the bacon its pink colour. An old book says that mature brine mixture gives a better cure than new brine, so it must have been the practice to save the brine from one pig to the next.

I once sampled a sweet-cured ham which was prepared by some students doing a home economy course. It was delicious, although the students did not like it. The ham was rubbed three times on consecutive days, then immersed in a special pickle. This was made by boiling up 4 gallons of water with 8 lbs black treacle, 2 lbs salt, 4 lbs brown sugar and 2 oz black pepper. This was simmered for twenty minutes and poured over the ham when cold. When the ham came out, it was left to dry slowly.

Home cured bacon is ready to eat in about a month, but it will keep for a long time.

Smoking. Smoking is done after curing as an optional extra. I don't think many people do it at home these days, but if anyone does I would be interested to hear about it. I gather from the recipe book that you can cheat by pouring essence of smoke over your cured ham. This is apparently crude pyroligneous acid.

William Cobbett was keen on smoking bacon. He did it in the chimney and although he bewails the scarcity of proper farmhouses with chimneys to smoke in, he says that most neighbourhoods will have one. I doubt whether this is still the case. His provisions are, no rain to come down the chimney to wet the bacon, and not too near the fire to be cooked - or to turn the fat rancid. The smoke must be wood-smoke and fir is not suitable. Cobbett says a month in the chimney will do. Before it goes in you will coat it with bran or sawdust, to keep it clean.

Sausages. Home made sausages are so much better to eat than bought ones (which are full of preservatives) that they deserve a mention. English sausages,

once described by a judge as 'bags of mystery', contain a lot of bread to make the meat go further and of course they are made to be fried. They will not keep - especially home made ones without preservative - but they will freeze well and thaw quickly for an unexpected meal if you freeze them in small packages.
Here is a basic recipe.

2lbs lean pork plus 1lb fat pork
1oz salt
1lb breadcrumbs
½oz pepper, pinch of nutmeg and mace
¼oz sage.

Mix breadcrumbs with meat, and mince all together. Add seasoning and put through the mincer again. This is the meat, which can be frozen as it is. To make it into sausages, you can get synthetic sausage skins - try you friendly local butcher. Old-timers used to use the pigs intestine, scraped and washed, as the skin, tied off at intervals; but if your pig is slaughtered at the abbatoir you won't get the intestines back, unless you take it away and dress it yourself.
The Kenwood electric mixer has an attachment for making saus ages.

Continental Sausages. These are entirely different; you see them hanging up in the delicatessen, and they will hang for months in your kitchen as a kind of affluent decoration. It seems a pity that the English cottagers never took up this idea because they are ready to eat without cooking and are very useful to have about.

The meat in these sausages is raw, preserved by the salt, spices and garlic, but over time it mellows and is delicious. Once I got really hooked on a Polish variation of this that a friend brought in when she had a Polish visitor.

The mixture of lean fat pork is the same as before, but no bread is added. Some people mince the meat and dice the pork fat finely, so that it shows up as little white squares.

Ingredients

3lbs minced meat and fat
1 glass wine or vinegar
3 cloves garlic
spices - e.g. paprika
brown sugar
1 tablespoon salt
pepper to taste
pinch saltpetre

The sausage is put into the casing - perhaps the pig's large intestine - and hung up in the kitchen; the optimum temperature is 60 deg. F.. You sometimes see these sausages with crossed strings round them. This is because they shrink with storage and may need to be laced up tight. They can be smoked for 24 hours in the chimney, if you have one of William Cobbett's proper chimneys to do it in. Sausage like this will keep for months - if it is not eaten first.

Raising a Pork Pie. If you kill a pig, somebody is bound to want a technical discussion on pork pies - 'raising'is the term used. Home made pork pie is a delicacy. In case the recipe turns out to be a closely-guarded secret, here is Mrs Beetons. She has a normal sized one using 5lbs lard and 14lbs flour for the crust, but I will spare you that. Here is her 'mode' for Little Raised Pork Pies.

Ingredients

2lbs flour
½lb butter
½lb mutton suet
4lbs neck of pork
1 dessertspoon pwd. sage
salt and pepper

Make sure the flour is dry and mince the suet. Melt the butter in a saucepan, add the flour, suet and a little salt. Mix it to a stiff paste, and keep it warm with a cloth over it until ready to use. Chop the pork into very small pieces and season it. To make a raised pork pie case: work the dough up the sides of a jar whilst it is still warm. This is important,otherwise the fat sets and the pastry becomes brittle. Bake in a pre set oven at 350deg. F. or Mark 4 for approximately 1 hour.

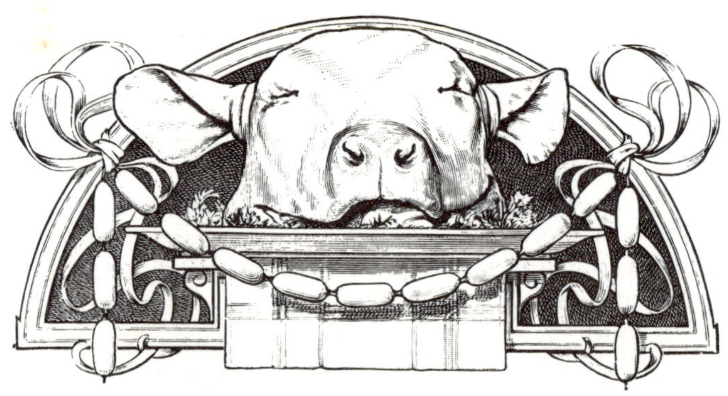

Making Leather

Most people leave the skin on their pigs and cook them in it, scoring it with a sharp knife before cooking to make carving easier. But there is not much food value in the skin, so it can be removed and cured for use as leather. Pigskin is traditionally used for things like gloves, wallets and saddles.

In the process of curing, the skin is chemically treated, often by tannin (tannic acid) to preserve it. This is an extremely old process, probably one of the first skills learned by mankind. It was not known until the nineteenth century how the process worked. Oak bark was the traditional tanning material, but other barks have been found to work as well.

There are other ways of curing. Formalin is sometimes used commercially, and this produces white 'doeskin' leather. Fats and oils will make a 'chamois' leather like that used for cleaning windows, which goes hard when dry but softens when wet. Wash leathers are usually made from a slice when the skin is split.

Smoking skins to cure them was a method used by the Red Indians, who also had a way of curing with the brains of an animal, which are apparently always the right size to cure the skin. The ancient Egyptians used alum and salt. This is called 'tawing' and is used for the production of kid leather.

The skin is composed of gelatine which enters into chemical combination with the tannic acid. The skin actually has three layers and only the middle layer or dermis is used to make leather. The outer layer or epidermis consists of dead cells, follicles, fat, sweat glands and blood vessels. This is removed, and so is the inner layer of flesh and fat.

Preparation of Skins

First the skin must be washed, perhaps in a dilute solution of carbolic acid. It should be taken in hand while fresh and not allowed to start decaying.

The next job is to remove the hair, but this may have

already been done because the normal process when a pig is killed is to scald the skin and scrape off the hair immediately. If it has not, boiling water should be poured over the skin and then it can be scraped, or it can be soaked in water to which lime has been added. A solution of 1 gallon quicklime to 10 gallons cold water can be used. The skins are soaked in this for 12 hours, after which they are washed with soap and water. If the lime is difficult to get rid of it can be neutralised with salt.

After this the skins are soaked in water containing pigeon's or dog's dung – at least they were, traditionally. Modern curers tend to keep quiet about this, but in the nineteenth century it was asserted that 'no dressing leather is ever submitted to bark without going through this process'.

After soaking in lime the hair should be loose and ready for scraping off. A fairly sharp instrument is used for pigskin, but be careful not to cut the skin. The inside is also scraped, a process known as fleshing. To prepare for this skins are sometimes soaked in a mixture of l part alcohol to 2 parts benzine in order to remove the grease.

After this the skins were put into tan pits with layers of oak bark between them, and then flooded with water and left for months with just an occasional stirring. Home tanners use the bark of many young trees; alder, birch,

chestnut, spruce, poplar and willow. After this the leather should be softened by pulling about over a blunt edged blade. Finishing the skin means taking off the roughness and this can be done with oil and stretching.

Rawhide can be useful round the holding and is far simpler to make. It is untanned leather and all that happens is that after dehairing and fleshing the skins are stretched on a frame and dry slowly, being pulled occasionally. It dries very hard and needs a sharp knife to cut it. It can be used for ropes if braided, for bindings, thongs and so on, and also for soling shoes. Rawhide can be greased to keep it flexible.

11 postscript

We come now to the consideration of our pigs in their surroundings. How do they fit in to backyarding and the countrside in general?

In my opinion, no holding is really complete without a pig. They are as we have seen, depositories for unwanted food scraps, efficient turners of the soil and fine producers of meat and manure. Pigs and dairy animals go very well together; nothing puts a 'bloom' on a pig's coat quicker than milk.

In a wider context, pigs have their place. You may want to keep a healthy outdoor pig of one of the older breeds. By doing so, you could be helping the breed to survive.

It is a melancholy fact that breeds of livestock which somehow don't fit in with the streamlined factory food production methods of the moment are being forgotten. Numbers are dwindling; over twenty separate breeds of domestic livestock have actually become extinct in the last thirty years.

Individuals can do little on their own, but those who feel concern about this state of affairs can join the Rare Breeds Survival Trust. This society takes an active part in preserving rare breeds for the future, for several reasons. Most of all they are worried about the loss of genetic potential when a breed is lost; who knows what traits we may need in the future to put vigour back into our in-bred stock? Who knows how things will change when the oil runs out?

Among the many examples of practical help which the Trust can give is advice on where to find, say, Gloucester Old Spot or Tamworth pigs. The journal of the RBST is called the Ark and is a mine of information about the old breeds. It is issued once a month to members.

While on the subject of conservation, I would like to mention another organisation dedicated to reversing regrettable trends. Much damage to wildlife has been done by the grubbing out of hedges, the draining of wet areas and the use of poisonous sprays. For years, conservationists have bewailed these things; but now they have begun to get involved with the people who actually farm the land.

The Farming and Wildlife Advisory Group contains farmers and conservationists, working together. They are concerned with the art of the possible. They would like farming and wild-life to exist side by side; their message is simple, but important to every land user, and that includes backyarders with an acre or two.

The idea is that everyone can help in a small way to provide cover for small animals and birds, possibly in a corner that is too awkward or small to be cultivated intensively. Cover is the main need and with this provided, wild things can usually look after themselves. Not long ago I visited a very large and modern farm, belonging to an agricultural college. It was surprising how much had been done there by planting trees in corners and in hedgerows, to improve the look of the place and to make it a better place for wild animals.

Bringing this down to backyard level, it means that you will be doing something positive if you leave the bushes in the corner of a field. We have been conditioned, some of us, into extreme tidiness; but weeds have their place, after all. Hedges are very important, because they are lines of communication, the corridors down which wild creatures move.

If you keep pigs on a small patch, keeping a good balance with the environment may be just a bit tricky. Overstocking is bad for the animals and it is also bad for the land; erosion can follow, and loss of soil structure. Pigs can easily make a desert for you, if you allow it.

The answer is to keep, if you can, a variety of stock and to consider the needs of the land. Get the animals off the fields in winter if the weather is wet or the soil is heavy. And remember the pig ring; this will save your grass from destruction. A copper ring, inserted into the nose with the proper pliers, is a great help in keeping outdoor pig activities under control.

reference

There are useful books, now out of print, published during and just after the war and dealing with keeping pigs. One is *Teach Yourself Good Pig Keeping*, revised edition 1958.

Back copies of the magazine *The Smallholder*, now defunct, would be very useful if you could find them.

Recent books about pigs are strictly commercial, based on a factory approach, but they may contain ideas which would be useful to you. They also give a glimpse of agribusiness at work. Here are a few.

Pig Husbandry. John Luscombe. Farming Press 1972.
This is practical and allows that there is a place for outdoor pigs.

Practical Pig Nutrition. Whittemore and Elsley. Farming Press 1976. In spite of the title this is very deep water; not really much use to backyarders, but very up-to-date.

Any Fool Can Be A Pig Farmer. J. Robertson. Farming Press 1975. Amusing personal account which gives the low-down on pig behaviour.

Practical Pig Production. Keith Thornton. Farming Press.

Pig Feeding. MAFF leaflet. HMSO.

Pig Farming is a monthly magazine published by Farming Press, Fenton House, Wharfedale Road, Ipswich, Suffolk. They will send you a free specimen copy on request; it is geared to the industry, but there are useful articles sometimes.

Buildings

Pig Housing. David Sainsbury. Farming Press 1972.

Build Your Own Farm Buildings. Frank Henderson. Farming Press - a classic.

Farm Buildings Pocket Book. MAFF. HMSO.

Health

Pig Diseases. H. Belschner. Angus and Robertson. 2nd Ed 1972. An Australian book, very detailed.

Pig Farmers' Vet Book. N. Barron. Farming Press.

TV Vet Book For Pig Farmers. Farming Press. 3rd Ed 1973. Easy to follow, many photographs.

Herbal Handbook For Farm and Stable. J. de B. Levy. Faber and Faber 1975. A lot about herbal treatments for sheep and although she does not deal with pigs specifically, I have found useful advice on treating pigs. A very good one for backyarders.

General Books

The Complete Book of Self Sufficiency. John Seymour. Faber 1976.

Fat of the Land. John Seymour. Faber. His earliest and best.

The Countryside Explained. John Seymour. Faber 1977.

Brave New Victuals. Elspeth Huxley. Chatto and Windus. Lowdown on bought food.

The Living Earth and the Haughley Experiment. E.B. Balfour. Faber reprint 1975. The principles of organic husbandry.

Cottage Economy. William Cobbett. Facsimile reprint by Landsmens Bookshop 1975.

The Farming Ladder. E. Henderson. Faber. Good sound principles.

Farmers of Forty Centuries. King. Cape 1926. (out of print). Organic sense.

The Owner Built Homestead Ken Kern. Prism Press.

The Owner Built Home Ken Kern. Prism Press.

Farming and Wildlife. Ed Barber. RSPB

Soil Association booklets—they are all useful.

Farming Organically.
Smallholder Harvest.
Self Sufficient Smallholding.
Use Your Weeds.

Agricultural Notebook. McConnell. Iliffe Books. Facts and figures.

Observers Book of Farm Animals L. Alderson. Warne.

Where to get supplies

Self Sufficiency and Smallholding Supplies
The Old Palace,
Priory Road,
Wells,
Somerset,
England.
All the things that are hard to get; hand tools, dairy equipment, small grinders, rennet etc.

Small Scale Supplies
Widdington,
Saffron Walden,
Essex.

"Everything for the self supporter". Books, greenhouses, cheese moulds, poultry equipment etc.

Dalton Supplies Ltd.
Nettlebed,
Oxfordshire,
England.

Eartags, weighbands for estimating pig weights, kicking bars for erring cows, etc.

Thompson and Morgan.
London Road,
Ipswich,
Suffolk.

Seeds including TREE seeds, unusual sorts; backyarding equipment and fish farming gear – good catalogue with information on growing.

Cade Horticultural Products Ltd.
Streetfield Farm,
Cade Street,
Heathfield,
Sussex.

Small amounts of farm seeds.

Rossendale,
25 Lumb,
Rossendale,
Lancashire.

Fencing – electrical mains.

Useful Addresses

The National Pig Breeders Association
49 Clarendon Road,
Watford,
Herts.

This is the organisation to put you in touch with sources of pigs, perhaps even in your own locality. They quote the following breeds: Berkshire, British Saddleback, Duroc, Gloucester Old Spot, Hampshire, Large Black, Pietrain, Large White, Middle White, Tamworth, Welsh. Perhaps the old breeds are making a comeback!

Messrs. Fergusons of New Pitsligo Ltd.
Cowbog,
New Pitsligo,
Aberdeenshire.

These people are breeders of hybrids designed for their own system of keeping pigs, which is very simple and allows the dry sows to go out to grass. They claim to have produced breeding stock with 'wide environmental tolerance'.

The Rare Breeds Survival Trust
c/o The Ark,
Winkleigh,
Devon.

The Soil Association,
Walnut Tree Manor,
Haughley,
Stowmarket,
Suffolk.

Advice on organic gardening and farming and a Journal.

The Henry Doubleday Research Association,
Convent Lane,
Bocking,
Braintree, Essex.